STTR

An Assessment of the Small Business Technology Transfer Program

Committee on Capitalizing on Science, Technology, and Innovation:
An Assessment of the Small Business Innovation Research Program—Phase II

Board on Science, Technology, and Economic Policy

Policy and Global Affairs

The National Academies of
SCIENCES · ENGINEERING · MEDICINE

THE NATIONAL ACADEMIES PRESS
Washington, D.C.
www.nap.edu

THE NATIONAL ACADEMIES PRESS 500 Fifth Street, NW Washington, DC 20001

This material is based upon work supported by the National Science Foundation under Grant No. DMI-0221736; by NASA under award No. NNX07AJ53G; by the National Institutes of Health under contract number HHSN263201200074I, Task Order HHSN26300029; by the Department of Energy under award number DE-PI0000010/DE-DT0006163; and by the Department of Defense under contract/order number HQ0034-10-D-0003-0001.

Any opinions, findings, and conclusions or recommendations expressed in this material are those of the author(s) and do not necessarily reflect the views of any organization or agency that provided support for the project.

International Standard Book Number-13: 978-0-309-37961-8
International Standard Book Number-10: 0-309-37961-X

Additional copies of this report are available for sale from the National Academies Press, 500 Fifth Street, NW, Keck 360, Washington, DC 20001; (800) 624-6242 or (202) 334-3313; http://www.nap.edu.

Printed in the United States of America

Suggested citation: National Academies of Sciences, Engineering, and Medicine. 2016. *STTR: An Assessment of the Small Business Technology Transfer Program.* Washington, DC: The National Academies Press.

The National Academies of

SCIENCES • ENGINEERING • MEDICINE

The **National Academy of Sciences** was established in 1863 by an Act of Congress, signed by President Lincoln, as a private, nongovernmental institution to advise the nation on issues related to science and technology. Members are elected by their peers for outstanding contributions to research. Dr. Ralph J. Cicerone is president.

The **National Academy of Engineering** was established in 1964 under the charter of the National Academy of Sciences to bring the practices of engineering to advising the nation. Members are elected by their peers for extraordinary contributions to engineering. Dr. C. D. Mote, Jr., is president.

The **National Academy of Medicine** (formerly the Institute of Medicine) was established in 1970 under the charter of the National Academy of Sciences to advise the nation on medical and health issues. Members are elected by their peers for distinguished contributions to medicine and health. Dr. Victor J. Dzau is president.

The three Academies work together as the **National Academies of Sciences, Engineering, and Medicine** to provide independent, objective analysis and advice to the nation and conduct other activities to solve complex problems and inform public policy decisions. The Academies also encourage education and research, recognize outstanding contributions to knowledge, and increase public understanding in matters of science, engineering, and medicine.

Learn more about the National Academies of Sciences, Engineering, and Medicine at **www.national-academies.org.**

Committee on Capitalizing on Science, Technology, and Innovation: An Assessment of the Small Business Innovation Research Program—Phase II

Jacques S. Gansler (NAE), *Chair*
Roger C. Lipitz Chair in Public Policy and Private Enterprise
Director of the Center for Public Policy and Private Enterprise
School of Public Policy
University of Maryland

David Audretsch
Distinguished Professor
Ameritech Chair
of Economic Development
Director of the Institute
for Development Strategies
Indiana University

Gene Banucci
Executive Chairman, ret.
ATMI, Inc.
(Member: 6/26/2009-4/23/2014)

Thomas J. Bond
Grant and Proposal Director
Association for Manufacturing
Technology
(Member: 6/26/2009-5/21/2014)

Michael Borrus
Founding General Partner
XSeed Capital

J. Michael Brick
Vice President and Co-Director
of Survey Methods
Westat

Gail H. Cassell (NAM)
Senior Lecturer
Department of Global Health
and Social Medicine
Harvard Medical School

M. Christina Gabriel
President
University Energy Partnership

Charles E. Kolb (NAE)
President and Chief Executive
Officer
Aerodyne Research, Inc.

Virginia Lesser
Professor of Statistics
Department of Statistics
Director, Survey Research Center
Oregon State University

Henry Linsert, Jr.
Chairman and CEO
Columbia Biosciences Corporation

W. Clark McFadden II
Senior Counsel
Orrick, Herrington
& Sutcliffe, LLP

Duncan T. Moore (NAE)
Vice Provost for Entrepreneurship
Rudolf and Hilda Kingslake
Professor of Optical Engineering
The Institute of Optics
University of Rochester

continued

Linda Powers
Managing Director
Toucan Capital Corporation
(Member: 6/26/2009-10/13/2011)

Donald Siegel
Dean and Professor
School of Business
University at Albany, SUNY

Jeffrey E. Sohl
Professor and Director of the Center for Venture Research
Peter T. Paul College of Business and Economics
University of New Hampshire

Tyrone C. Taylor
President
Capitol Advisors on Technology, LLC

John P. Walsh
Professor of Public Policy
School of Public Policy
Georgia Institute of Technology

Patrick H. Windham
Principal
Technology Policy International

Project Staff

Sujai J. Shivakumar
Study Director

David E. Dierksheide
Program Officer

Karolina E. Konarzewska
Program Coordinator
(through September 2015)

Gail E. Cohen
Board Director

Frederic A. Lestina
Senior Program Assistant

Natacha R. Montgomery
Senior Program Assistant
(through October 2015)

For the National Academies of Sciences, Engineering, and Medicine, this project was overseen by the Board on Science, Technology, and Economic Policy (STEP), a standing board established in 1991. The mandate of the Board on Science, Technology, and Economic Policy is to advise federal, state, and local governments and inform the public about economic and related public policies to promote the creation, diffusion, and application of new scientific and technical knowledge to enhance the productivity and competitiveness of the U.S. economy and foster economic prosperity for all Americans. The STEP Board and its committees marshal research and the expertise of scholars, industrial managers, investors, and former public officials in a wide range of policy areas that affect the speed and direction of scientific and technological change and their contributions to the growth of the U.S. and global economies. Results are communicated through reports, conferences, workshops, briefings, and electronic media subject to the procedures of the Academies to ensure their authoritativeness, independence, and objectivity. The members and staff of the STEP Board* are listed below:

* As of December 2015.

continued

Laura D'Andrea Tyson
Professor of Business
Administration & Economics
Director, Institute for Business
& Social Impact
Haas Business & Public Policy
Group
University of California, Berkeley

Jay Walker
Chairman
Patent Properties, Inc.

STEP Staff

Gail E. Cohen
Director

Paul T. Beaton
Senior Program Officer

Aqila A. Coulthurst
Associate Program Officer

David E. Dierksheide
Program Officer

Frederic A. Lestina
Senior Program Assistant

Natacha R. Montgomery
Senior Program Assistant
(thorough October 2015)

Sujai J. Shivakumar
Senior Program Officer

Preface

Today's knowledge economy is driven in large part by the nation's capacity to innovate. One of the defining features of the U.S. economy is a high level of entrepreneurial activity. Entrepreneurs in the United States see opportunities and are willing and able to assume risk to bring new welfare-enhancing, wealth-generating technologies to the market. Yet, although discoveries in areas such as genomics, bioinformatics, and nanotechnology present new opportunities, converting these discoveries into innovations for the market involves substantial challenges.[1] The American capacity for innovation can be strengthened by addressing the challenges faced by entrepreneurs. Public-private partnerships are one means to help entrepreneurs bring new ideas to market.

The Small Business Innovation Research (SBIR) program and the Small Business Technology Transfer (STTR) program together form one of the largest examples of U.S. public-private partnerships. An underlying tenet of these programs is that small businesses are a strong source of new ideas, and therefore economic growth, but that it is difficult to find financial support for these ideas in the early stages of their development. The SBIR program was established in 1982 to encourage small businesses to develop new processes and products and to provide quality research in support of the U.S. government's many missions. By involving qualified small businesses in the nation's research and development (R&D) effort, SBIR awards stimulate innovative technologies to help federal agencies meet their specific R&D needs in many areas, including health, the environment, and national defense. The STTR program was created in 1992 by the Small Business Research and Development Enhancement Act to expand joint venture opportunities for small businesses and nonprofit research institutions by requiring small business recipients to collaborate formally with a research institution.

---

[1]See L. M. Branscomb, K. P. Morse, M. J. Roberts, D. Boville, *Managing Technical Risk: Understanding Private Sector Decision Making on Early Stage Technology Based Projects,* Gaithersburg, MD: National Institute of Standards and Technology, 2000.

In the SBIR Reauthorization Act of 2000, Congress tasked the National Research Council (NRC)[2] with undertaking a "comprehensive study of how the SBIR program has stimulated technological innovation and used small businesses to meet federal research and development needs" and with recommending further improvements to the program.[3] In the first round of this study, an expert committee prepared a series of reports from 2004 to 2009 on the SBIR program at the Department of Defense (DoD), National Institutes of Health (NIH), National Aeronautics and Space Administration (NASA), Department of Energy (DoE), and National Science Foundation (NSF)—the five agencies responsible for 96 percent of the program's operations.[4] When reauthorizing the SBIR and STTR programs in 2011, Congress expanded the study mandate to include a review of the STTR program.[5]

Building on the methodology and outcomes from the previous review of SBIR, this assessment is a part of a series of reports that examines topics of general policy interest that emerged during the first round as well as topics of specific interest to individual agencies administering SBIR and STTR. In addition to this report, which reviews the STTR program across the five agencies identified above, the results of the assessment will be published in reports of agency-specific and program-wide findings on the SBIR and STTR programs to be submitted to the contracting agencies and Congress. In partial fulfillment of these objectives, this volume presents the committee's review of the STTR program.[6]

ACKNOWLEDGMENTS

On behalf of the National Academies of Sciences, Engineering, and Medicine, we express our appreciation for and recognition of the valuable insights and close cooperation extended by agency staff, the survey respondents, and case study interviewees, among others. The committee gives particular thanks to its lead researcher, Robin Gaster of Innovation Competitions LLC, and to Peter Grunwald of Grunwald Associates LLC, which conducted the surveys and described the results presented in this volume. Rosalie Ruegg of TIA Consulting provided valuable assistance in revising the draft report in light of comments received from reviewers. The presentation of the report has also been enhanced by the diligent copyediting of Nancy Tuvesson. David Dierksheide of

[2]Effective July 1, 2015, the institution is called the National Academies of Sciences, Engineering, and Medicine. References in this report to the National Research Council are used in an historic context identifying programs prior to July 1.

[3]See the SBIR Reauthorization Act of 2000 (H.R. 5667, Section 108).

[4]For the overview report, see National Research Council, *An Assessment of the SBIR Program,* Washington, DC: The National Academies Press, 2008. See also National Research Council, *An Assessment of the SBIR Program at the National Aeronautics and Space Administration,* Washington, DC: The National Academies Press, 2009. The committee also prepared reports on the SBIR program at DoD, DoE, NIH, and NSF.

[5]SBIR/STTR Reauthorization Act of 2011, P.L. 112-81, December 31, 2011.

[6]The formal Statement of Task is presented in Chapter 1 of this report.

the staff of the Board on Science, Technology, and Economic Policy is especially recognized for his dedication and important contributions to the operation of this study and the preparation of this report.

Acknowledgment of Reviewers

This report has been reviewed in draft form by individuals chosen for their diverse perspectives and technical expertise, in accordance with procedures approved by the National Academies of Sciences, Engineering, and Medicine's Report Review Committee. The purpose of this independent review is to provide candid and critical comments that will assist the institution in making its published report as sound as possible and to ensure that the report meets institutional standards for objectivity, evidence, and responsiveness to the study charge. The review comments and draft manuscript remain confidential to protect the integrity of the process.

We wish to thank the following individuals for their review of this report: Wendy Baldwin, Population Reference Bureau; Robert Barnhill, Arizona State University; Glenn Firebaugh, Pennsylvania State University; Kenneth Gall, Duke University; Terry Grimm, Niowave, Inc.; Karl Koster, Massachusetts Institute of Technology; Jane Muir, University of Florida; Philip Neches, Teradata Corporation; Richard Nelson, Columbia University; Phillip Phan, Johns Hopkins University; Carl Ray, NASA (Retired); Marcia Rieke, University of Arizona; John Younger, University of Michigan; and Lorel Wisniewski, Independent Consultant.

Although the reviewers listed above have provided many constructive comments and suggestions, they were not asked to endorse the conclusions or recommendations, nor did they see the final draft of the report before its release. The review of this report was overseen by Edwin Przybylowicz, Eastman Kodak Company (retired) and Irwin Feller, Pennsylvania State University. Appointed by the Academies, they were responsible for making certain that an independent examination of this report was carried out in accordance with institutional procedures and that all review comments were carefully considered. Responsibility for the final content of this report rests entirely with the authoring committee and the institution.

Jacques S. Gansler Sujai J. Shivakumar

Contents

Summary

The Small Business Technology Transfer (STTR) program, partnered with the Small Business Innovation Research (SBIR) program, provides innovation awards that expand joint venture opportunities for small businesses and nonprofit research institutions.

Created in 1982 through the Small Business Innovation Development Act,[1] SBIR remains the nation's largest innovation program for small businesses. SBIR offers competitive awards to support the development and commercialization of innovative technologies by small private-sector businesses. At the same time, SBIR provides government agencies with technical and scientific solutions that address their different missions.

Created in 1992 by the Small Business Research and Development Enhancement Act of 1992, STTR seeks to bridge the gap between basic science and commercialization of resulting innovations.[2] Under the STTR program, a small business receiving an award must collaborate formally with a research institution (RI).

Both the SBIR and STTR programs share a three-phase structure:

- Phase I provides limited funding (up to $100,000 prior to the 2011 reauthorization and up to $150,000 thereafter) for feasibility studies.
- Phase II provides more substantial funding for further research and development (typically up to $750,000 prior to the 2011 reauthorization and $1 million thereafter).[3]

---

[1]Small Business Innovation Development Act of 1982, P.L. 97-219, S. 881, July 22, 1982.
[2]Small Business Research and Development Enhancement Act, P.L. 102-564, S. 2941, October 28, 1992.
[3]All resource and time constraints imposed by the program are somewhat flexible and are addressed by different agencies in different ways. For example, the National Institutes of Health—and to a

- Phase III reflects commercialization without providing access to any additional SBIR/STTR funding, although funding from other federal government accounts is permitted.

The SBIR program has four congressionally mandated goals: (1) stimulate technological innovation, (2) use small business to meet federal research and development (R&D) needs, (3) foster and encourage participation by minority and disadvantaged persons in technological innovation, and (4) increase private-sector commercialization derived from federal research and development.[4]

By comparison, the statutory objective for the STTR program is "to stimulate a partnership of ideas and technologies between innovative small business concerns (SBCs) and Research Institutions through Federally-funded research or research and development (R/R&D)."[5] However, the Small Business Administration (SBA) web site also lists additional STTR objectives that are largely aligned with the SBIR program: (1) stimulate technological innovation, (2) foster technology transfer through cooperative R&D between small businesses and research institutions, and (3) increase private-sector commercialization of innovations derived from federal R&D.

STTR is administered by the Department of Defense (DoD), National Institutes of Health (NIH), Department of Energy (DoE), National Science Foundation (NSF), and National Aeronautics and Space Administration (NASA). Each of these research agencies has the flexibility to administer SBIR and STTR in line with its own unique mission needs.[6]

Although the SBIR and STTR programs have similar objectives, there are important differences. Under SBIR the principal investigator (PI) must be primarily employed with the small business concern (SBC) at the time of award and for the duration of the project period. Under STTR, however, the PI "may be primarily employed by either the small business concern or the collaborating non-profit research institution at the time of award and for the duration of the project period."[7]

The programs also differ in that STTR requires the SBC to formally collaborate with a nonprofit research institution. Research partnerships are permitted under the SBIR program, but the partnering research institution can complete no more than one-third of the Phase I work and no more than one-half of the Phase II work. In contrast, "Under STTR, the small business must perform at least 40 percent of the work and the research institution must perform

---

much lesser degree the Department of Defense—have provided awards that are much larger than the standard amounts.

[4]Small Business Innovation Development Act of 1982, P.L. 97-219, S. 881, July 22, 1982.

[5]Small Business Administration, STTR Policy Directive, February 2014, p. 3.

[6]An Academies committee commended this flexibility in a 2008 assessment of the SBIR program. See Finding C, National Research Council, *An Assessment of the SBIR Program*, Washington, DC: The National Academies Press, 2008, p. 59.

[7]See, for example, the NIH web site at https://sbir.nih.gov/about/critical, accessed on July 9, 2015.

at least 30 percent. The remaining 30 percent may be … [completed by] the small business concern, the collaborating non-profit research institution, or an additional third party."[8]

For fiscal year (FY) 2015, funding was approximately $2.17 billion for SBIR, compared to approximately $263 million for STTR. While STTR is dwarfed in size by SBIR (agency budgets for SBIR are seven or eight times larger than those for the STTR), the Small Business Administration views STTR as a vehicle to expand funding opportunities in the federal innovation R&D arena. In particular, STTR is expected to combine the strengths of research institutions and small firms by introducing entrepreneurial skills to high-tech research efforts. This design has sought to assist the transfer of technologies and products from the laboratory to the marketplace.

CALL FOR ASSESSMENT

Adopting several recommendations from a 2008 National Research Council (NRC)[9] report, Congress reauthorized the SBIR and STTR programs in December 2011 for an additional 6 years. As a part of this reauthorization, Congress called for further studies by the National Academies of Sciences, Engineering, and Medicine of the SBIR and STTR programs.

The findings and recommendations of the Academies committee are summarized below. They are based on a complement of quantitative and qualitative tools including surveys, case studies of award recipients, agency data, public workshops, and discussions with agency managers. The methodology is described in Chapter 1 and Appendix A of this report. Appendix C displays the survey, and Appendix E presents 11 illustrative case studies. These case studies humanize and illuminate the findings, and provide a richer feel for the subject and the people involved.

The survey data presented in this report cover 1,400 STTR awards made by the five study agencies during the period FY1998-2010, which represents the preliminary survey population. Awards made in FY2010 would only have started to generate commercial products by the time of the survey in FY2014; therefore, following established practice from prior Academies and Government Accountability Office (GAO) surveys, no awards were surveyed after FY2014. Given the time period covered it is not surprising that many points of contact could not be reached: 807 of the 1,400 contacts were not reachable, leaving a population of 593. In all, 292 questionnaires were answered, generating a response rate of 20.9 percent for the preliminary population and 49.2 percent for the population.

---

[8]Ibid.

[9]Effective July 1, 2015, the institution is called the National Academies of Sciences, Engineering, and Medicine. References in this report to the National Research Council or NRC are used in an historic context identifying programs prior to July 1.

KEY FINDINGS

General conclusions about the STTR program must be viewed with caution. STTR programs are managed and operated differently by each agency and in some cases differently by separate components within DoD and NIH. Therefore, individual agencies will need to consider the findings and recommendations provided herein within the specific context of their own programs. Not all findings will be relevant to all agencies. See Chapter 6 for the full list of findings.

A. STTR is meeting its congressional objective of fostering cooperation between small business concerns and research institutions, and does so in some respects to an extent that SBIR does not.

1. Overall, the university connection is much deeper and richer for STTR awards than for SBIR, and STTR addresses its congressional mandate to stimulate partnerships between small business concerns and research institutions to an extent that SBIR does not.
2. STTR projects generate wider and deeper linkages between small businesses and research institutions than do SBIR projects, according to data from the Academies 2011-2014 Survey.

B. Perspectives on STTR use and management vary by agency. Some see it as a link between basic research and acquisition programs; others see STTR as having similar objectives to SBIR and therefore operate the two programs in tandem.

1. Program managers at NASA and DoD (in particular the Army and Navy) see STTR as filling a gap between basic research and acquisition programs.
2. Program managers at NIH, NSF, and DoE do not see an additional value in the STTR program. They see STTR as having similar objectives to SBIR and therefore operate the two programs in tandem.

C. To a considerable extent, STTR fosters private-sector commercialization of innovations derived from federal R&D.

D. The participation of women and minorities in the STTR program is low and not actively fostered.

1. Data from the Academies 2011-2014 Survey indicate that the participation of women in the STTR program is low. Survey respondents reported that woman-owned firms accounted for 8 percent of all STTR Phase II firms.

2. Data from the Academies 2011-2014 Survey indicate that the participation of Black, Hispanic, and Native Americans in the STTR program is extremely low.
3. The SBA definition of socially or economically disadvantaged groups is inadequate to reflect congressional objectives. Data reported by the agencies obscures the extremely low level of participation from other disadvantaged groups by including Asian Americans. The Academies survey found that only 1.1 percent of respondents were from Black- and Hispanic-owned firms respectively, and 0.4 percent was from Native American-owned firms.

E. STTR is aligned with agency missions and the take-up of technologies within acquisition agencies.

F. STTR awards require a formal partnership between the small business concern and the research institution, but they each can have different interests and needs. This creates unique challenges within the STTR program.

1. In particular, research institutions see the development and widespread dissemination of technical knowledge as part of their core mission. In contrast, small businesses see the commercialization of knowledge as a priority, which will likely require steps to limit the ability of others to use technical information, through the use of either trade secrets or patents.
2. As university faculty participate in commercial activities outside the research institution, university administrators often seek to ensure that a dividing line exists between research inside the university and activities outside.
3. Research institutions have varied views on and approaches to licensing of university intellectual property (IP).
4. The bureaucracy at research institutions can be challenging. Research institutions are big organizations, typically with large overhead rates, and the transfer of technology often is not seen as a core part of their mission. Unless there is a defined path to partnership, negotiating with research institutions can take considerable time and resources for a small business.

G. Small business concerns in general see STTR as more onerous to use and thus less attractive than SBIR, in part because STTR awards require a formal partnership between the small business and the research institution.

H. STTR supports the development of innovative companies.

KEY RECOMMENDATIONS

The committee finds that STTR meets the specific congressional objective of increasing the linkages between small business concerns and research institutions. To encourage more small businesses to collaborate with research institutions, the committee recommends:

A. The five sponsoring agencies should address the following factors that may be discouraging some small businesses and research institutions from collaborating in the STTR and SBIR programs:

1. STTR Program
    - **Finding alternative templates for royalties and licensing agreements.** The complexity and variation in intellectual property terms and conditions among universities and laboratories can cause delays in developing contractual agreements between the research institution and small business concerns. The potential partners in an STTR award should consult leading research institutions to learn what templates for royalty and licensing schemes have proven to be most effective and might be adapted for their project. Many different schemes have been used and should be reviewed in the context of the potential project and its participants. One example, adopted at the University of Minnesota, offers a standard option along with an alternate "open negotiation" option that could be a useful template for some projects.
    - **Resolving unique challenges of cooperation.** Given the highly flexible nature of the STTR program, the sponsoring agencies should consider seeking SBA authority to act in special circumstances to protect participants from the effects of unexpected delays or related problems with contract agreements or deliverables.
    - **Maintaining a distinct strategy for STTR**. Each sponsoring agency should seek ways to ensure that the STTR program plays an identifiable role in the agency's R&D strategy that differs from that played by the SBIR program. A focus on projects with earlier technology readiness levels might be part of this strategic distinction.

2. SBIR Program
    - **Relaxing the small business employment requirement**. Research institutions with personnel who seek to serve as Principal Investigators on SBIR awards while retaining their full-time positions might be allowed—under exceptional circumstances—to

seek a waiver of the SBIR 51 percent small business employment requirement.

- **Reporting on waiver requests.** If any waivers are to be considered, the sponsoring agencies should develop an appropriate mechanism for addressing these special requests and should report on the number of waiver requests and the number granted, as part of their annual program reporting.

3. The overall impact of these proposed changes should be evaluated in future assessments of the SBIR and STTR programs to determine if they have been effective in strengthening the collaboration between small business concerns and research institutions in the STTR and SBIR programs.

B. SBA should change its definitions to address congressional intent with regard to minorities.

1. SBA should act immediately to change its definitions to ensure that efforts in this area are focused on activities that meet congressional intent.
2. SBA should also require that agencies collect data—and report annually—on the participation of each SBA subgroup in the SBIR and STTR programs.

1

Introduction

Small businesses are an important driver of innovation and economic growth in the United States.[1] Despite the challenges of changing global environments and the impacts of the 2008 financial crisis and subsequent recession, innovative small businesses continue to develop and commercialize new products for the market, improving the health and welfare of Americans while strengthening the nation's security and competitiveness.[2]

Created in 1982 through the Small Business Innovation Development Act,[3] the Small Business Innovation Research (SBIR) program remains the nation's largest innovation program for small businesses. The SBIR program offers competitive awards to support the development and commercialization of innovative technologies by small private-sector businesses.[4] At the same time,

---

[1]See Z. Acs and D. Audretsch, "Innovation in large and small firms: An empirical analysis," *The American Economic Review*, 78(4):678-690, 1988. See also Z. Acs and D. Audretsch, *Innovation and Small Firms*, Cambridge, MA: The MIT Press, 1991; E. Stam and K. Wennberg, "The roles of R&D in new firm growth," *Small Business Economics,* 33:77-89, 2009; E. Fischer and A.R. Reuber, "Support for rapid-growth firms: A comparison of the views of founders, government policymakers, and private sector resource providers," *Journal of Small Business Management*, 41(4):346-365, 2003; M. Henrekson and D. Johansson, "Competencies and institutions fostering high-growth firms," *Foundations and Trends in Entrepreneurship*, 5(1):1-80, 2009.

[2]See D. Archibugi, A. Filippetti, and M. Frenz, "Economic crisis and innovation: Is destruction prevailing over accumulation?" *Research Policy*, 42(2):303-314, 2013. The authors show that "the 2008 economic crisis severely reduced the short-term willingness of firms to invest in innovation" and also that it "led to a concentration of innovative activities within a small group of fast growing new firms and those firms already highly innovative before the crisis." They conclude that "the companies in pursuit of more explorative strategies towards new product and market developments are those to cope better with the crisis."

[3]Small Business Innovation Development Act of 1982, P.L. 97-219, July 22, 1982.

[4]SBIR awards can be made as grants or as contracts. Grants do not require the awardee to provide an agreed deliverable (for contracts this is often a prototype at the end of Phase II). Contracts are also governed by federal contracting regulations, which are considerably more onerous from the small business perspective. Historically, all Department of Defense (DoD) and National Aeronautics and Space Administration (NASA) awards have been contracts; all National Science Foundation (NSF)

the program provides government agencies with technical and scientific solutions that address their different missions.

Seeking to bridge the gap between basic science and commercialization of resulting innovations, the Small Business Technology Transfer (STTR) program, created in 1992 by the Small Business Research and Development Enhancement Act of 1992,[5] sought to expand joint venture opportunities for small businesses and nonprofit research institutions. Under the STTR program, a small business receiving an award must collaborate formally with a research institution.

Both the SBIR and STTR programs share a three-phase structure:

- Phase I provides limited funding (up to $100,000 prior to the 2011 reauthorization and up to $150,000 thereafter) for feasibility studies.
- Phase II provides more substantial funding for further research and development (typically up to $750,000 prior to 2012 and $1 million after the 2011 reauthorization).[6]
- Phase III reflects commercialization without providing access to any additional SBIR/STTR funding, although funding from other federal government accounts is permitted.

The SBIR program has four congressionally mandated goals: (1) stimulate technological innovation, (2) use small business to meet federal research and development (R&D) needs, (3) foster and encourage participation by minority and disadvantaged persons in technological innovation, and (4) increase private-sector commercialization derived from federal research and development.[7]

In comparison, the statutory objective for the STTR program is "to stimulate a partnership of ideas and technologies between innovative small business concerns (SBCs) and Research Institutions through Federally-funded research or research and development (R/R&D)."[8] However, the Small Business Administration (SBA) web site also gives additional objectives that are largely aligned with the SBIR program: (1) stimulate technological innovation, (2) foster technology transfer through cooperative R&D between small businesses and research institutions, and (3) increase private-sector commercialization of innovations derived from federal R&D.[9]

---

and most National Institutes of Health (NIH) awards have been grants, and the Department of Energy (DoE) has used both vehicles.

[5]Small Business Research and Development Enhancement Act, P.L. 102-564, S. 2941, October 28, 1992.

[6]All resource and time constraints imposed by the program are somewhat flexible and are addressed by different agencies in different ways. For example, NIH and to a much lesser degree DoD have provided awards that are much larger than the standard amounts, and NIH has a tradition of offering no-cost extensions to allow for completion of work on an extended timeline.

[7]Small Business Innovation Development Act of 1982, P.L. 97-219, S. 881, July 22, 1982.

[8]SBA, STTR Policy Directive, February 2014, p. 3.

[9]SBA, https://www.sbir.gov/about/about-sttr, accessed on November 27, 2015.

Each of the research agencies has sought to pursue these goals in administering its SBIR and STTR programs, utilizing the administrative flexibility to address its own unique mission needs.[10]

Although the SBIR and STTR programs have similar objectives, there are important differences. Under SBIR, the Principal Investigator (PI) must be primarily employed with the small business concern (SBC) at the time of award and for the duration of the project period, but primary employment is not stipulated in the STTR Program. Under STTR, the principal investigator "may be primarily employed by either the small business concern or the collaborating non-profit research institution at the time of award and for the duration of the project period."[11]

The programs also differ in that STTR requires the small business concern to formally collaborate with a nonprofit research institution. Research partnerships are permitted under the SBIR program, but the partnering research institution can complete no more than one-third of the Phase I work and no more than one-half of the Phase II work. In contrast, "Under STTR, the small business must perform at least 40 percent of the work and the research institution must perform at least 30 percent. The remaining 30 percent may be … [completed by] the small business concern, the collaborating non-profit research institution, or an additional third party."[12]

Over time, through a series of reauthorizations, the legislation enabling SBIR and STTR has required federal agencies with extramural R&D budgets in excess of $100 million to set aside a growing share of their budgets for the SBIR program and those with extramural R&D budgets in excess of $1 billion to set aside a growing share of their budgets for the STTR program (see Table 1-1). By fiscal year (FY)2012, the 11 federal agencies listed in Table 1-2 that administer SBIR and STTR programs were disbursing $2.4 billion dollars a year.[13] Five agencies administer greater than 96 percent of SBIR/STTR funds: Department of Defense (DoD), Department of Health and Human Services (HHS; particularly NIH), National Aeronautics and Space Administration (NASA), National Science Foundation (NSF), and Department of Energy (DoE). Aggregate award amounts for the five largest agencies for FY2015 are provided in Table 1-1. STTR is administered only at these five agencies.

Although STTR is dwarfed in size by SBIR (agency budgets for SBIR are seven or eight times larger than those of STTR), SBA views it as a vehicle to expand funding opportunities in the federal innovation R&D arena. In particular, STTR was designed to combine the strengths of research institutions

---

[10]The committee commended this flexibility in its 2008 assessment of the SBIR program. See Finding C, National Research Council, *An Assessment of the SBIR Program*, Washington, DC: The National Academies Press, 2008, p. 59.

[11]See, for example, the NIH web site at https://sbir.nih.gov/about/critical, accessed July 9, 2015.

[12]Ibid.

[13]SBA, SBIR/STTR annual report, http://www.sbir.gov/, accessed July 2015. FY2012 is the most recent year for which SBA publishes comparative data across agencies.

TABLE 1-1 SBIR/STTR Funding by the Five Principal Funding Agencies, FY 2015

	Funding (Thousands of Dollars)	
Agency	STTR	SBIR
DoD	118,840	1,070,758
HHS	86,933	656,480
DoE	23,464	169,797
NASA	18,531	139,184
NSF	15,452	131,305
Total	263,220	2,167,524

SOURCE: Small Business Administration, SBIR/STTR Annual Report, http://www.sbir.gov, accessed October 20, 2015.

TABLE 1-2 Agencies Participating in the SBIR and STTR Programs in 2015

Agency	SBIR Participant	STTR Participant
Department of Agriculture	X	
Department of Commerce	X	
Department of Defense	X	X
Department of Education	X	
Department of Energy	X	X
Department of Health and Human Services	X	X
Department of Homeland Security	X	
Department of Transportation	X	
Environmental Protection Agency	X	
National Aeronautics and Space Administration	X	X
National Science Foundation	X	X

SOURCE: Small Business Administration.

and small firms by introducing entrepreneurial skills to high-tech research efforts. This design has sought to assist the transfer of technologies and products from the laboratory to the marketplace.

HISTORY AND STRUCTURE OF THE SBIR AND STTR PROGRAMS[14]

During the 1980s, the perceived decline in U.S. competitiveness due to Japanese industrial growth in sectors traditionally dominated by U.S. firms—

[14]Parts of this section are based on the Academies' previous report on the NIH SBIR program—National Research Council, *An Assessment of the SBIR Program at the National Institutes of Health*, Washington, DC: The National Academies Press, 2009.

autos, steel, and semiconductors—led to concerns about future economic growth in the United States.[15] A key concern was the perceived failure of American industry "to translate its research prowess into commercial advantage."[16] Although the United States enjoyed dominance in basic research—much of which was federally funded—applying this research to the development of innovative products and technologies remained a challenge. As the great corporate laboratories of the post-war period were buffeted by change, new models such as the cooperative model utilized by some Japanese keiretsu seemed to offer greater sources of dynamism and more competitive firms.[17]

At the same time, new evidence emerged to indicate that small businesses were an increasingly important source of both innovation and job creation.[18] This evidence reinforced recommendations from federal commissions, dating back to the 1960s, that federal R&D funding should provide more support for innovative small businesses (which was opposed by traditional recipients of government R&D funding).[19]

Early-stage financial support for innovative technology-based small businesses for developing high-risk technologies with commercial promise was first advanced by Roland Tibbetts at NSF. In 1976, Mr. Tibbetts advocated shifting some NSF funding for this purpose. NSF adopted this initiative first, and after a period of analysis and discussion, the Reagan administration supported its expansion across the federal government. Congress then passed the Small Business Innovation Research Development Act of 1982, which established the SBIR program.

---

[15]See J. Alic, "Evaluating competitiveness at the office of technology assessment," *Technology in Society*, 9(1):1-17, 1987, for a review of how these issues emerged and evolved within the context of a series of analyses at a Congressional agency.

[16]D.C. Mowery, "America's industrial resurgence (?): An overview," in D.C. Mowery, ed., *U.S. Industry in 2000: Studies in Competitive Performance*, Washington, DC: National Academy Press, 1999, p. 1. Other studies highlighting poor economic performance in the 1980s include M.L. Dertouzos et al., *Made in America: The MIT Commission on Industrial Productivity*, Cambridge, MA: The MIT Press, 1989; and O. Eckstein, *DRI Report on U.S. Manufacturing Industries*, New York: McGraw Hill, 1984.

[17]Keiretsu is "the name given to a form of corporate structure in which a number of organisations link together, usually by taking small stakes in each other and usually as a result of having a close business relationship, often as suppliers to each other." See *The Economist*, "Keiretsu," October 16, 2009. http://www.economist.com/node/14299720.

[18]For an alternate view, see S.J. Davis, J. Haltiwanger, and S. Schuh, Small Business and Job Creation: Dissecting the Myth and Reassessing the Facts, Working Paper No. 4492, Cambridge, MA: National Bureau of Economic Research, 1993. Evaluating the empirical basis for conventional claims about the job-creating prowess of small businesses, the authors find inter alia that conventional wisdom about the job-creating prowess of small business rests on misleading interpretations of the data. According to Per Davidsson, these methodological fallacies, however, "ha[ve] not had a major influence on the empirically based conclusion that small firms are over-represented in job creation." See P. Davidsson, "Methodological concerns in the estimation of job creation in different firm size classes," Working Paper, Jönköping International Business School, 1996.

[19]For an overview of the origins and history of the SBIR program, see G. Brown and J. Turner, "The federal role in small business research," *Issues in Science and Technology*, Summer 1999, pp. 51-58.

Initially, the SBIR program required agencies with extramural R&D budgets in excess of $100 million[20] to set aside 0.2 percent of their funds for SBIR. Program funding totaled $45 million in the program's first year of operation (1983). Over the next 6 years, the set-aside grew to 1.25 percent.[21]

SBIR Reauthorizations of 1992 and 2000

The SBIR program approached reauthorization in 1992 amidst continued worries about the ability of U.S. firms to commercialize inventions (see Box 1-1). Finding that "U.S. technological performance is challenged less in the creation of new technologies than in their commercialization and adoption," the Academies recommended an increase in SBIR funding as a means to improve the economy's ability to adopt and commercialize new technologies.[22]

The Small Business Research and Development Enhancement Act (P.L. 102-564) reauthorized the SBIR program until September 30, 2000, and doubled the set-aside rate to 2.5 percent. The legislation also more strongly emphasized the need for commercialization of SBIR-funded

BOX 1-1
Commercialization Language from 1992 SBIR Reauthorization

Phase II "awards shall be made based on the scientific and technical merit and feasibility of the proposals, as evidenced by the first phase, considering, among other things, the proposal's commercial potential, as evidenced by

(i) the small business concern's record of successfully commercializing SBIR or other research;
(ii) the existence of second phase funding commitments from private sector or non-SBIR funding sources;
(iii) the existence of third phase, follow-on commitments for the subject of the research; and
(iv) the presence of other indicators of the commercial potential of the idea."

SOURCE: P.L. 102-564, October 28, 1992.

---

[20]That is, those agencies spending more than $100 million on research conducted outside agency labs.
[21]Additional information regarding SBIR's legislative history can be accessed from the Library of Congress. See http://thomas.loc.gov/cgi-bin/bdquery/z?d097:SN00881:@@@L.
[22]See National Research Council, *The Government Role in Civilian Technology: Building a New Alliance*, Washington, DC: National Academy Press, 1992, p. 29.

technologies.[23] Legislative language explicitly highlighted commercial potential as a criterion for awarding SBIR contracts and grants.

At the same time, Congress expanded the SBIR program's purposes to "emphasize the program's goal of increasing private sector commercialization developed through federal research and development and to improve the federal government's dissemination of information concerning the small business innovation, particularly with regard to woman-owned business concerns and by socially and economically disadvantaged small business concerns."[24]

The Small Business Reauthorization Act of 2000 (P.L. 106-554) extended the SBIR program until September 30, 2008. It also called for a National Research Council (NRC)[25] assessment of the program's broader impacts, including those on employment, health, national security, and national competitiveness.[26]

STTR Establishment and Reauthorizations

Established by the Small Business Technology Transfer Act of 1992 (P.L. 102-564, Title II), the STTR program was reauthorized until the year 2001 by the Small Business Reauthorization Act of 1997 (P.L. 105-135) and reauthorized again until September 30, 2009, by the Small Business Technology Transfer Program Reauthorization Act of 2001 (P.L. 107-50).[27]

As explained below, the SBIR/STTR Reauthorization Act of 2011 included a number of changes to the SBIR-STTR programs, including increases in the set-asides over the next 6 years and expanded eligibility for STTR awardees to take part in technical assistance programs.

---

[23]Small Business Research and Development Enhancement Act, P.L. 102-564, S. 2941, October 28, 1992. See also R. Archibald and D. Finifter, "Evaluation of the Department of Defense Small Business Innovation Research program and the Fast Track Initiative: A balanced approach," in National Research Council, *The Small Business Innovation Research Program: An Assessment of the Department of Defense Fast Track Initiative,* Washington, DC: National Academy Press, 2000, pp. 211-250.

[24]Small Business Research and Development Enhancement Act, P.L. 102-564, S. 2941, October 28, 1992.

[25]Effective July 1, 2015, the institution is called the National Academies of Sciences, Engineering, and Medicine. References in this report to the National Research Council or NRC are used in an historic context identifying programs prior to July 1.

[26]The current assessment is congruent with the Government Performance and Results Act (GPRA) of 1993: http://govinfo.library.unt.edu/npr/library/misc/s20.html. As characterized by the Government Accountability Office (GAO), GPRA seeks to shift the focus of government decision making and accountability away from a preoccupation with the activities that are undertaken—such as grants dispensed or inspections made—to the results of those activities. See http://www.gao.gov/new.items/gpra/gpra.htm.

[27]See https://www.sbir.gov/about, accessed on December 3, 2015.

The 2011 SBIR/STTR Reauthorization

The anticipated 2008 reauthorization was delayed in large part by a disagreement between long-time program participants and their advocates in the small business community and proponents of expanded access for venture-backed firms, particularly in biotechnology where proponents argued that the standard path to commercial success includes venture funding at some point.[28] Other issues were also difficult to resolve, but the conflict over participation of venture-backed companies dominated the process[29] following an administrative decision to exclude these firms more systematically.[30]

After a much extended discussion, passage of the National Defense Act of December 2011 reauthorized the SBIR/STTR programs through FY2017.[31] The new law maintained much of the core structure of both programs but made some important changes, which were to be implemented via the SBA's subsequent Policy Guidance.[32]

The eventual compromise on the venture funding issue allowed (but did not require) agencies to award up to 25 percent of their SBIR grants or contracts (at NIH, DoE, and NSF) or 15 percent (at the other awarding agencies) to firms that benefit from private, venture capital investment. It is too early in the implementation process to gauge the impact of this change.[33]

The reauthorization made changes to the SBIR program that were recommended in prior Academies reports.[34] These included the following:

- Increased award size limits
- Expanded program size
- Enhanced agency flexibility—for example, for Phase I awardees from other agencies to be eligible for Phase II awards or to award an additional Phase II
- Improved incentives for the utilization of SBIR technologies in agency acquisition programs
- Explicit requirements for better connecting prime contractors with SBIR awardees

---

[28]D.C. Specht, "Recent SBIR extension debate reveals venture capital influence," *Procurement Law,* 45:1, 2009.

[29]W.H. Schacht, "The Small Business Innovation Research (SBIR) program: Reauthorization efforts," Congressional Research Service, Library of Congress, 2008.

[30]A. Bouchie, "Increasing number of companies found ineligible for SBIR funding," *Nature Biotechnology,* 21(10):1121-1122, 2003.

[31]SBIR/STTR Reauthorization Act of 2011, P.L. 112-81, December 31, 2011.

[32]See SBA post, S. Greene, "Implementing the SBIR and STTR Reauthorizations: Our Plan of Attack," http://www.sbir.gov/news/implementing-sbir-and-sttr-reauthorization-our-plan-attack, accessed February 21, 2012.

[33]See National Research Council, *Venture Capital and the NIH SBIR Program*, Washington, DC: The National Academies Press, 2009.

[34]See Appendix B for a list of the major changes to the SBIR program resulting from the 2011 Reauthorization Act.

- Substantial emphasis on developing a more data-driven culture, which has led to several major reforms, including the following:
    - adding numerous areas of expanded reporting
    - extending the Academies' evaluation
    - adding further evaluation, such as by the Government Accountability Office and Comptroller General
    - tasking the SBA with creating a unified platform for the collection of data

- Expanded management resources (through provisions permitting use of up to 3 percent of program funds for [defined] management purposes)
- Expanded commercialization support (through provisions providing companies with direct access to commercialization support funding and through approval of the approaches piloted in Commercialization Pilot Programs)
- Options for agencies to add flexibility by developing other pilot programs—for example, to allow awardees to skip Phase I and apply for a Phase II award directly or for NIH to support a new Phase 0 pilot program.

The reauthorization also made changes that were not mentioned in previous reports of the Academies. These included the following:

- Expansion of the STTR program
- Limitations on agency flexibility—particularly in the provision of larger awards
- Introduction of commercialization benchmarks for companies, which must be met if companies are to remain in the program. These benchmarks are to be established by each agency.

Other clauses of the legislation affect operational issues, such as the definition of specific terms (such as "Phase III"), continued and expanded evaluation by the Academies, mandated reports from the Comptroller General on combating fraud and abuse within the SBIR program, and protection of small firms' intellectual property within the program.

PREVIOUS RESEARCH ON SBIR

Studies pre-dating the Academies' first-round assessment, most notably by the General Accounting Office and the SBA, focused only on specific aspects or components of the SBIR/STTR programs.[35] In addition, prior to the first-round assessment, there had been few internal assessments of agency SBIR/STTR programs. The academic literature on SBIR was also limited,[36] except for an assessment in the 1990s by Joshua Lerner of the Harvard Business School, who found "that SBIR awardees grew significantly faster than a matched set of firms over a ten-year period."[37]

To help fill this assessment gap for the SBIR/STTR programs, the NRC's Committee for Government-Industry Partnerships for the Development of New Technologies (GIP, which preceded the NRC's first-round congressionally mandated study of the SBIR program) convened a workshop to discuss the SBIR program's history and rationale, review existing research, and identify areas for further research and program improvements.[38] In addition, in its report on the SBIR Fast Track Initiative at DoD, the GIP committee found that the SBIR program contributed to mission goals by funding "valuable innovative projects."[39] It concluded that a significant number of these projects would not have been undertaken absent SBIR funding[40] and that DoD's Fast Track Initiative encouraged the commercialization of new technologies[41] and the entry of new firms into the program.[42] The GIP committee also found that the SBIR program improved both the development and utilization of human capital and the diffusion of technological knowledge.[43] Case studies provided some evidence that the knowledge and human capital generated by the SBIR program have positive economic value, which spills over into other firms through the movement of people and ideas.[44] Furthermore, by providing a validation of

---

[35]An important step in the evaluation of the program has been to identify existing evaluations of the program. These include U.S. General Accounting Office, *Federal Research: Small Business Innovation Research Shows Success But Can Be Strengthened*, Washington, DC: U.S. General Accounting Office, 1992; and U.S. General Accounting Office, *Evaluation of Small Business Innovation Can Be Strengthened*, Washington, DC: U.S. General Accounting Office, 1999. There is also a 1999 unpublished SBA study on the commercialization of SBIR Phase II awards from 1983 to 1993 among non-DoD agencies.

[36]Early examples of evaluations of the SBIR program include S. Myers, R. L. Stern, and M. L. Rorke, *A Study of the Small Business Innovation Research Program*, Lake Forest, IL: Mohawk Research Corporation, 1983; and Price Waterhouse, *Survey of Small High-tech Businesses Shows Federal SBIR Awards Spurring Job Growth, Commercial Sales*, Washington, DC: Small Business High Technology Institute, 1985.

[37]See J. Lerner, "The government as venture capitalist: The long-run effects of the SBIR program," *Journal of Business*, 72(3), 1999.

[38] See National Research Council, *The Small Business Innovation Research Program: Challenges and Opportunities*, Washington, DC: National Academy Press, 1999.

[39]National Research Council, *The Small Business Innovation Research Program: An Assessment of the DoD SBIR Fast Track Initiative*, Washington, DC: National Academy Press, p. 32.

[40]Ibid., p. 32.

[41]Ibid., p. 33.

[42]Ibid., p. 34.

[43]Ibid., p. 33.

[44]Ibid., p. 33.

promising new technologies, SBIR awards encourage further private-sector investment in an award-winning firm's technology.[45]

FIRST-ROUND ASSESSMENT OF THE SBIR PROGRAM

The 2000 SBIR reauthorization mandated that the NRC complete a comprehensive assessment of the SBIR program.[46] The separate assessment of the SBIR programs at DoD, NIH, NASA, NSF, and DoE began in 2002 and was conducted in three steps. As a first step, the committee authoring this study developed a research methodology[47] and gathered information about the program by convening workshops where officials at the relevant federal agencies described their program operations, challenges, and accomplishments. These meetings highlighted the important differences in agency goals, practices, and evaluations. They also served to describe the evaluation challenges that arise from the diversity in program objectives and practices.[48]

The committee implemented the research methodology during the second step. As set out in the methodology, multiple data collection modalities were deployed. These included the first large-scale survey of SBIR award recipients. Case studies of a wide variety of SBIR firms were also developed. The committee then evaluated the results and developed the findings and recommendations presented for improving the effectiveness of the SBIR program.

During the third step, the committee reported on the SBIR program through a series of publications in 2008-2010: five individual volumes on the major funding agencies and an additional overview volume titled *An Assessment of the SBIR Program.*[49] Together, these reports provided the first detailed and comprehensive review of the SBIR program and, as noted above, served as an important input into SBIR reauthorization prior to December 2011 (see Box 1-2).

CURRENT, SECOND-ROUND STUDY: CHALLENGES AND OPPORTUNITIES

The first-round study of the SBIR program found that the program was, overall, "sound in concept and effective in practice."[50] The current study,

[45]Ibid., p. 33.
[46]SBIR Reauthorization Act of 2000, P.L. 106-554, Appendix I-H.R. 5667, Section 108.
[47]National Research Council, *An Assessment of the Small Business Innovation Research Program: Project Methodology*, Washington, DC: The National Academies Press, 2004.
[48]Adapted from National Research Council, *SBIR: Program Diversity and Assessment Challenges,* Washington, DC: The National Academies Press, 2004.
[49]National Research Council, *An Assessment of the SBIR Program.*
[50]Ibid., p. 54.

BOX 1-2
The First-Round Assessment of the Small Business Innovation Research (SBIR) Program

Mandated by Congress in the 2000 reauthorization of the SBIR program, the National Research Council's first-round SBIR assessment reviewed the SBIR programs at the Department of Defense, National Institutes of Health, National Aeronautics and Space Administration, Department of Energy, and National Science Foundation. In addition to the reports on the SBIR program at each agency and a report on the program methodology, the study resulted in a summary of a symposium on program diversity and assessment challenges, a summary of a symposium on the challenges in commercializing SBIR-funded technologies, two reports on special topics, as well as the committee's summary report, *An Assessment of the SBIR Program*. In all, 11 study reports were published by the National Academies Press:

An Assessment of the Small Business Innovation Research Program: Project Methodology (2004)
SBIR—Program Diversity and Assessment Challenges: Report of a Symposium (2004)
SBIR and the Phase III Challenge of Commercialization: Report of a Symposium (2007)
An Assessment of the SBIR Program at the National Science Foundation (2007)
An Assessment of the SBIR Program at the Department of Defense (2009)
An Assessment of the SBIR Program at the Department of Energy (2008)
An Assessment of the SBIR Program (2008)
An Assessment of the SBIR Program at the National Aeronautics and Space Administration (2009)
An Assessment of the SBIR Program at the National Institutes of Health (2009)
Venture Funding and the NIH SBIR Program (2009)
Revisiting the Department of Defense SBIR Fast Track Initiative (2009)

described in the Statement of Task in Box 1-3, provides a second snapshot to measure the SBIR program's progress against its legislative goals. Importantly, the second-round study also includes an assessment of the STTR program.

This volume on the STTR program partially addresses this Statement of Task. It is supplemented by a number of workshops and other publications that assess agency SBIR programs (See Box 1-2). For example, workshops were convened on the participation of women and minorities in the SBIR-STTR programs (February 2013), the evolving role of university participation in the

BOX 1-3
Statement of Task

In accordance with H.R. 5667, Sec. 108, enacted in Public Law 106-554, as amended by H.R. 1540, Sec. 5137, enacted in Public Law 112-81, the National Research Council is to review the Small Business Innovation Research and Small Business Technology Transfer (SBIR/STTR) programs at the Department of Defense, the National Institutes of Health, the National Aeronautics and Space Administration, the Department of Energy, and the National Science Foundation. Building on the outcomes from the first-round study, this second-round study is to examine both topics of general policy interest that emerged during the first-round study and topics of specific interest to individual agencies.

Drawing on the methodology developed in the previous study, an ad hoc committee will issue a revised survey, revisit case studies, and develop additional cases, thereby providing a second snapshot to measure the program's progress against its legislative goals. The committee will prepare one consensus report on the SBIR program at each of the five agencies, providing a second review of the operation of the program, analyzing new topics, and identifying accomplishments, emerging challenges, and possible policy solutions. The committee will prepare an additional consensus report focused on the STTR Program at all five agencies. The agency reports will include agency-specific and program-wide findings on the SBIR and STTR programs to submit to the contracting agencies and Congress.

Although each agency report will be tailored to the needs of that agency, all reports will, where appropriate:

1. Review institutional initiatives and structural elements contributing to programmatic success, including gap funding mechanisms such as applying Phase II-plus awards more broadly to address agency needs and operations and streamlining the application process.
2. Explore methods to encourage the participation of minorities and women in SBIR and STTR.
3. Identify best practice in university-industry partnering and synergies with the two programs.
4. Document the role of complementary state and federal programs.
5. Assess the efficacy of post-award commercialization programs.

In partial fulfillment of this Statement of Task, this volume presents the committee's review of the operation of the STTR program.

programs (February 2014), the relationship between state innovation programs and the SBIR program (October 2014), the STTR program (May 2015), and the economics of entrepreneurship in relation to the SBIR program (June 2015). The committee published a report on *Innovation, Diversity, and Success in the SBIR/STTR Programs*, based on the 2013 workshop. Relevant to this publication, the committee also convened a workshop on the STTR program on May 1, 2015.

Information on which to assess the STTR program has been drawn from the Academies 2011 and 2014 surveys, which is described in detail in Appendix A, company case studies profiled in Appendix E, discussions with university technology transfer officials, a series of ongoing discussions and conversations with agency officials, and the workshop convened by the committee on the STTR program in Washington, DC on May 1, 2015 (see Box 1-4). A guide to the contents, data sources, and organization of this report can be found at the conclusion of this chapter.

STUDY METHODOLOGY

The SBIR-STTR programs are unique in terms of scale and mission focus. In addition, the evidence suggests that there are no truly comparable

BOX 1-4
The STTR Workshop

The committee convened a May 2015 workshop on the STTR program, drawing on the experiences of program managers from the five participating agencies as well as entrepreneurs from high-technology small businesses with experience in SBIR and STTR. See Appendix I for the workshop agenda. Issues explored at the workshop included:

- How do DoD, NIH, DoE, NSF, and NASA run their STTR programs? How is STTR operationally different from SBIR? What do we know about program outcomes?
- What are the advantages of collaboration between small business and research institutions, including national laboratories?
- What are the main barriers to meeting Congressional objectives more fully?
- What program adjustments would better support commercialization?
- Are there aspects of the program that make it less attractive? Could they be addressed?
- Can the program generate better data on both process and outcomes and use those data to fine-tune program management?
- In what other ways can STTR be improved?

This report draws in the insights gained at this workshop.

programs in the United States, and those in other countries operate in such different ways that their relevance is limited.[51] Thus, it is difficult to identify programs comparable to SBIR-STTR against which to benchmark their results.

Assessing the SBIR-STTR programs is challenging given the diversity of the agencies involved. At DoD and NASA, SBIR-STTR awards are primarily designed to generate tools and capabilities for agency use. At the other study agencies they are instead explicitly designed to generate technologies that will be adopted outside the agency, primarily in the private sector. Thus commercialization success cannot be measured across agencies only in terms of project outputs sold to the agency.

The SBIR-STTR programs are also highly decentralized at some agencies. At NIH for example, the SBIR-STTR program office within the NIH Office of Extramural Programs sets policy and provides critical cross-agency communication flows, as well as links the program to outside stakeholders, but award funding is separately determined by each Institute or Center (IC). ICs take different views of the program and use different approaches to program management. Therefore, generalizations about the SBIR and STTR programs must be made with care.

Focus on Legislative Objectives

This study is focused on assessing the extent to which STTR is meeting its congressionally mandated objectives and on providing recommendations for further program improvements.[52] It provides assessment-based findings of the benefits and costs of STTR while seeking to improve the public's understanding of the program—including the differing perspectives of entrepreneurs, agency staff, research institutions, and other stakeholders—and makes recommendations to improve the program's effectiveness.

Definition Challenges

Commercialization offers practical and definitional challenges. As described in Chapter 5, several different definitions of commercialization can be used when discussing the SBIR-STTR programs. In fact, it is important to use more than one simple definition. For example, the percentage of funded projects that reach the marketplace is not the only measure of commercial success.

---

[51]See National Academies of Sciences, Engineering, and Medicine, Workshop on "Learning from Each Other: U.S. European Perspectives on Small Business Innovation Programs," Washington, DC, March 19, 2015.

[52]These limited objectives are consistent with the methodology developed by the committee. See National Research Council, *An Assessment of the Small Business Innovation Research Program: Project Methodology*.

In the private sector, commercial success over the long term requires profitability. However, in the short term, the path to successful commercialization can involve many different aspects of commercial activity, from product rollout to licensing to patenting to acquisition. Even during new product rollout, companies often do not generate immediate profits. This second-round assessment uses multiple metrics to address the question of commercialization (see Chapter 5).

Quantitative Assessment Methods

More practically, several issues relate to the application of quantitative assessment methods, including decisions about which kinds of program participants should be targeted for survey deployment, the number of responses that are appropriate, selection bias, nonresponse bias, the design and implementation of survey questionnaires, and the level of statistical evidence required for drawing conclusions in this case. These and other issues were discussed at a workshop and summarized in a 2004 report.[53] Also, as noted above, a peer-reviewed report on the study methodology completed by the first-round committee provided the baseline for the initial study and for follow-on studies—including this one.[54]

Survey Development

For the current study, a survey of SBIR and STTR award recipients was developed and deployed. The survey was based closely on previous surveys, particularly the 2005 Survey that focused exclusively on SBIR, but nonetheless included significant improvements.[55] The description of the survey and improvements, including a discussion of the survey outreach and response, are documented in Appendix A. Most notably, the survey development made an ambitious but ultimately unsuccessful effort to develop a comparison group to provide context and a benchmark for analyzing the results (this effort is also discussed in Appendix A).

The survey covered 1,400 STTR awards made by the five study agencies during the period FY1998-2010. There was a recorded point of contact for each of the 1,400 awards. This is the preliminary survey population. Awards made in FY2010 would only have started to generate commercial products by the time of the survey in FY2014, so following established practice from prior Academies and GAO surveys, no awards made after FY2010 were surveyed.

---

[53]National Research Council, *SBIR: Program Diversity and Assessment Challenges*.

[54]National Research Council, *An Assessment of the Small Business Innovation Research Program: Project Methodology*.

[55]The survey carried out as part of this study was administered in 2014, and the survey completed as part of the Academies' first-round assessment of the SBIR program was administered in 2005. In this volume all survey references are to the 2014 Survey unless noted otherwise.

Given the time period covered it is not surprising that many points of contact could not be reached. Of the 1,400 contacts comprising the preliminary survey population, 807 could not be reached, leaving an effective population of 593. Of these, 292 answered questionnaires, generating a response rate of 20.9 percent for the preliminary population and 49.3 percent for the effective population.

Appendix A provides a detailed discussion of the issues related to quantitative methodologies, as well as a review of potential biases. As a result of the relatively small response rate, there are significant limitations on the conclusions that can be drawn from this quantitative assessment, which is reflected in the wording of findings and recommendations (Chapter 6). At the same time, drawing on quantitative analysis is a crucial component of the overall study, reflective of the need to identify and assess outcomes that are found only by querying contacts in participating companies for individual STTR project awards.

A Complement of Approaches

Partly because of these limitations, the 2004 methodology report stressed the importance of utilizing a complement of research modalities, an approach that has been adopted here.[56] Although quantitative assessment represents the bedrock of our research and provides insights and evidence that could not be generated through any other modality, it is, in and of itself, insufficient to address the multiple questions posed in this analysis. Consequently, we undertook a series of additional activities:

- **Case studies.** We conducted in-depth case studies of 11 STTR recipients. These companies were geographically and demographically diverse, funded by different agencies, focused on different kinds of technologies, and at different stages of the company lifecycle. The case studies themselves are included as Appendix E.
- **Workshops.** We conducted workshops, including workshops to discuss the participation of women and minorities and the role of universities in the SBIR-STTR programs, as well as a workshop focused on the STTR program,[57] to allow stakeholders, agency staff, and academic experts to

[56]National Research Council, *An Assessment of the Small Business Innovation Research Program: Project Methodology*.

[57]Workshops convened by the committee as part of the overall analysis include *NASA Small Business Innovation Research Program Assessment: Second Phase Analysis*, January 28, 2010; *Early-Stage Capital in the United States: Moving Research Across the Valley of Death and the Role of SBIR*, April 16, 2010; *Early-Stage Capital for Innovation—SBIR: Beyond Phase II*, January 27, 2011; *NASA's SBIR Community: Opportunities and Challenges*, June 21, 2011; *Innovation, Diversity, and Success in the SBIR/STTR Programs*, February 7, 2013; *Commercializing University Research: The Role of SBIR and STTR*, February 5, 2014; *SBIR/STTR & the Role of State Programs*, October 7, 2014; *The Small Business Technology Transfer Program*, May 1, 2015, and the

provide insights into program operations and to identify issues that need to be addressed.

- **Analysis of agency data.** As appropriate, we analyzed and included data from the five STTR agencies that cover various aspects of program activities.
- **Open-ended responses from SBIR-STTR recipients.** For the first time, we collected textual responses in the survey. These comments are addressed in Chapter 4.
- **Agency consultations.** We engaged in discussions with staff at each of the five major agencies about the operation of their STTR programs and the challenges they face.
- **Literature review.** Since the start of our research in this area, a number of academic and policy papers have been published that address various aspects of the SBIR-STTR programs, many drawing from the survey and other data made available by our reviews. In addition, other organizations—such as the GAO—have reviewed specific parts of the SBIR-STTR programs. The committee has incorporated references to this work, where useful, into its analysis.

Data Sources and Limitations

Multiple research modalities are especially important because limitations still exist in the data collected for the SBIR-STTR programs.

In particular, there is no unified source of data on outcomes. Some agencies have significant data on outcomes—notably DoD and DoE. Other such as NIH and NASA are still developing their systems. Also, in some cases, such as NSF, the agency preferred not to provide outcomes data on privacy grounds.

Accordingly, the survey was used to provide quantitative insights into the program as a whole. While the limits of this methodology are described in detail in Appendix A, the survey provided important information about both outcomes and the views of recipients on various aspects of program operations and management. In particular, the survey allows for an analysis of the ways in which STTR connects research institutions and small businesses, data that are especially important in assessing the extent to which STTR meets its congressional objective

In short, within the limitations described, the study utilizes a complement of tools to ensure that a wide spectrum of perspectives and expertise is reflected in the findings and recommendations. Appendix A provides an overview of the methodological approaches, data sources, and survey tools used in this study.

Economics of Entrepreneurship, June 29, 2015. Each of these workshops was held in Washington, DC.

ORGANIZATION OF THE REPORT

The analysis and conclusions are organized as follows. Chapter 2 provides a review of program operations, describing the program in some detail and addressing a range of issues related to program management. Chapter 3 provides an overview of applications and awards, illustrating trends in the size of the program over time. Chapter 4 provides a qualitative assessment of the program, based on material drawn from interviews conducted with companies and the wider case studies included in Appendix E, as well as textual responses from survey respondents. Chapter 5 draws on the Academies survey to provide a quantitative assessment, covering the congressional objective, other SBA objectives, and the longer term impact of the program on recipient companies, as well as the role of women and minorities in the program. Chapter 6 provides the findings and recommendations from the study.

The report's appendixes provide additional information. Appendix A sets out an overview of the methodological approaches, data sources, and survey tools used in this assessment. Appendix B describes key changes to the SBIR and STTR programs from the 2011 reauthorization. Appendix C reproduces the Academies survey instrument. Appendix D lists the research institutions involved in STTR awards. Appendix E presents the case studies of selected firms with STTR awards. Appendix F is the data annex for Chapter 3, and Appendix G is the data annex for Chapter 5. Appendix H provides a glossary of acronyms used, Appendix I includes the agenda from the May 1, 2015, Academies STTR workshop, and Appendix J provides a list of references.

2

Program Management

The STTR program is operated by the five largest federal research agencies: the Department of Defense (DoD), Department of Energy (DoE), National Aeronautics and Space Agency (NASA), National Institutes of Health (NIH), and National Science Foundation (NSF). This chapter describes how each agency runs its program, addressing similarities and differences. It underscores that there are important differences between the *acquisition agencies* (DoD and NASA) who seek to purchase technologies developed using STTR, through contracts and deliverables, and the *grant agencies* (DoE, NIH, and NSF) who provide funding for projects that are aligned with the agency's mission, but aim for commercialization outside the agency and seek to acquire the technology for agency use only in few cases.

In recent years, DoD and NASA have worked hard to ensure that there is close alignment between the needs of the acquisition groups within the agency and the topics published for their SBIR and STTR programs.[1] While this effort is commendable, it also carries with it potential costs, notably that some of the more speculative research that is less aligned with specific acquisition programs is no longer being funded, and that earlier stage or higher risk research is also likely to be discouraged. In addition, the demands by the agency for funded technologies may be quite limited, while the agency specification may not closely match company market opportunities and may make it difficult for them to enter broader commercial markets with their project outputs.

The challenge of continuing to support high-risk high-reward research aligns with the opportunity provided by STTR to link small business concerns (SBC's) more closely with research institutions (RIs) (including universities,

[1]See National Research Council, *SBIR at the Department of Defense*, Washington, DC: The National Academies Press, 2014, and National Academies of Sciences, Engineering, and Medicine, *SBIR at the National Aeronautics and Space Administration*, forthcoming. Effective July 1, 2015, the institution is called the National Academies of Sciences, Engineering, and Medicine. References in this report to the National Research Council or NRC are used in an historic context identifying programs prior to July 1.

nonprofit research foundations, and national labs). Because staff and faculty at research institutions usually focus on earlier stage research, ensuring that there is a connection to these institutions and a pathway for the eventual commercialization of this technology is important.

As described in Chapter 4, there is also a range of perspectives on the value of STTR from different company participants, survey respondents, and agency staff. This chapter provides a description of how the different agencies actually operate their programs, underscoring the differences between the STTR program and the much larger SBIR program. This lays the groundwork for understanding where STTR adds value for the agencies.

STTR FUNDING

Following reauthorization in December 2011, Congress has mandated that the percentage of extramural research funding set aside by the research agencies for STTR should increase to the following percentages of budget:

- not less than 0.35 percent of such budget in fiscal years (FYs) 2012 and 2013;
- not less than 0.40 percent of such budget in fiscal years 2014 and 2015; and
- not less than 0.45 percent of such budget in fiscal year 2016 and each fiscal year after.[2]

DIFFERENTIATING SBIR AND STTR

Some of the differences between SBIR and STTR are mandated by the governing legislation and by Small Business Administration (SBA) policy guidance. Other differences emerge at individual agencies through program management and implementation.

Constraints on the Principal Investigator

Like SBIR, STTR legislation imposes requirements on the small business concern and the principal investigator (PI). However, under STTR, employment restrictions for the PI are in critical respects less burdensome than those for SBIR, where the PI must be at least 51 percent employed at the applicant company. In comparison, STTR permits the PI to remain primarily employed at the partner research institution (RI). This difference makes STTR

[2]SBA STTR Policy Directive, December 2014, p. 3.

much more attractive for faculty members who wish to retain their academic positions, although university regulations also play a role in this area.[3]

Some agencies also mandate level of effort. For example, at DoE SBIR/STTR PIs must devote for the duration of the project a minimum of 3 hours per week if it is in Phase I and 5 hours per week if it is in Phase II.[4] Some other agencies do not mandate such restrictions.

Requirements for Research Institution Participation

Under SBIR, there is no requirement that the research team include a research institution, and no requirement that any specified amount of funding flow to the RI. SBIR awardees are required to perform at least two-thirds of the research or analytical effort in house for Phase I and at least 50 percent for Phase II, and there are no constraints on the selection of subcontractors. Agencies may decide that "research and analytical effort" includes the entire award, or they may exclude some equipment and indirect costs.

Under STTR Phase I and II, at least 40 percent of the research or analytical effort must be performed by the applicant and at least 30 percent must be performed by a single RI. Such research institutions include universities and federal laboratories. The remaining 30 percent may be spent at either entity or on third-party goods or services.

While the small businesses must remain the responsible party, the application requires that the small business concern and the research institution enter into a formal partnership and reach a formal agreement on project-related IP. The relationship between the small business and the research institution may be quite different under STTR.

BOX 2-1
The Benefits of STTR Partnerships

Dr. Green of Physical Sciences Inc. observes that STTR cannot just be pass-through funding to the research institution. In a case study, presented in Appendix E of this report, he notes that STTR encourages each partner to work to their strength: "the research institution does research and education, and the industry partner does commercialization, and this structure is perfect for technology transition."

---

[3]DoE Phase I SBIR/STTR Funding Opportunity Announcement, DE-FOA-0001366, August 17, 2015, p. 17.

[4]Ibid. Applications for which the PI does not work greater than 50 percent time at the applicant company will automatically be considered as STTR applications at DoE.

Phase I Timelines

For some agencies (e.g., NASA), STTR provides for a longer timeline for Phase I completion, based on the rationale that organizing a collaboration with a research institution takes additional time. Most agencies (but not all, and some but not all components at DoD) permit 12 months to complete an STTR Phase I project, compared to 6 months for the basic SBIR Phase I project.[5] Phase II timelines are set at 2 years for both SBIR and STTR.

These broad requirements pertaining to principal investigator and research institution involvement in STTR are implemented across all of the agencies; however, as with SBIR, the program is managed differently at the different agencies, and even within the DoD components. The following section describes STTR activities at the five study agencies.

STTR AT THE FIVE STUDY AGENCIES

Department of Defense

For each phase at DoD, success rates for STTR are consistently higher than those for SBIR. The Phase I success rate is 16 percent for SBIR and 22 percent for STTR,[6] and the conversion rate from initial Phase I application to eventual Phase II award was 8 percent for SBIR and 11 percent for STTR. For both, the conversion rate to Phase III was 4 percent (see Figure 2-1).[7]

The primary research partners for DoD, by a wide margin, are universities (as opposed to other RIs). Eighty-eight percent of STTR Phase I awards and 83 percent of STTR Phase II awards have a university partner. STTR companies also partner with other small companies to a considerable extent (more than one-third of Phase II projects do so), which illustrates the notion that small companies are a part of a larger innovation ecology that serves DoD (see Table 2-1). To a much lesser extent, STTR companies also partner with large companies. Overall, research partners received about $34 million in FY 2014 from the DoD STTR program.

In 2013, 220 universities participated in STTR partnerships at DoD. Pennsylvania State University participated in the most projects—3.5 percent of all DoD STTR awards.

---

[5]DoD timing for SBIR Phase I awards is complicated by the adoption of Phase I options at some components, which provide additional time and resources, as well as by the Air Force use of a 9-month Phase I with no options.

[6]Data from the chart in Figure 2-1 do not match exactly with DoD awards and applications data in Chapter 3 because they are based on different time periods.

[7]To see the presentation that Figure 2-1 was drawn from, as well as other presentations and the webcast of the Academies May 1, 2015, STTR workshop, access at http://sites.nationalacademies.org/PGA/step/PGA_160863.

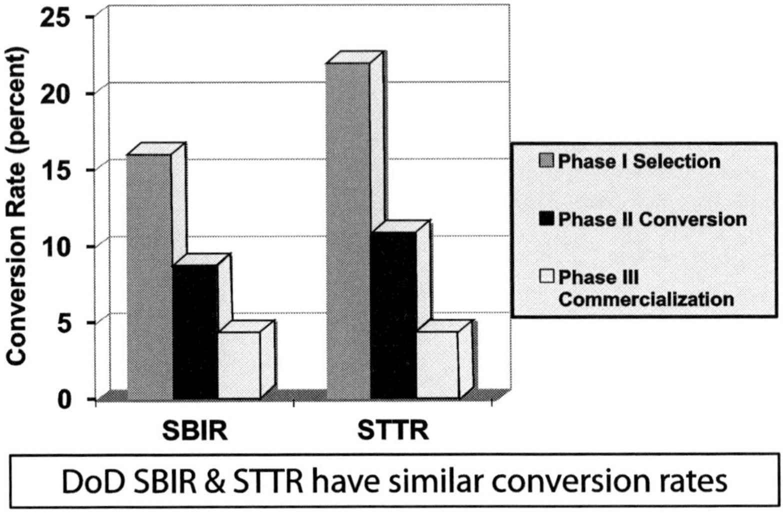

FIGURE 2-1 SBIR and STTR conversion rates at DoD.
SOURCE: Christopher Rinaldi, "STTR at DoD," presentation at the Academies STTR Workshop, May 1, 2015, p. 3. Access at http://sites.nationalacademies.org/PGA/step/PGA_160863.

TABLE 2-1 Research Partners for Phase I and Phase II STTR Projects at DoD

Phase I	Phase II
88% Universities	83%Universities
4% FFRDC	6% FFRDC
5% Nonprofit	4% Nonprofit
28% Small Businesses	37%Small Businesses
6% Large Businesses	6% Large Businesses
2% Other	5% Other

NOTE: FFRDC denotes Federal Funded Research and Development Corporations.
SOURCE: Christopher Rinaldi, "STTR at DoD," presentation at the Academies STTR Workshop, May 1, 2015.

Although research institutions (and particularly universities) participate in SBIR projects, they play a far less prominent role and conclude far fewer teaming agreements than in STTR projects (see Table 2-2).

Component Perspectives on STTR

The different components at DoD use STTR in different ways. Army STTR Program Manager Bradley Guay noted that more than one-half of Army

STTR topics in recent years were generated by PhD scientists at the Army Research Center and were not reviewed by potential customers within Army.[8] He emphasized that Army STTR topics were more research-oriented and that topic authors had to justify the use of STTR by explaining the importance of the connection with research institutions. It is notable that nonc of the Army Program Executive Offices formally participates in the Army STTR program and that more than one-third of topics originate in the Army Research Lab (see Figure 2-2). In contrast, all Navy topics are reviewed by program executive officers, who have acquisitions authority.

TABLE 2-2 Teaming Agreements in SBIR Awards at DoD

Phase I	Phase II
18% Universities	14% Universities
0.5%FFRDC	0.1%FFRDC
2% Nonprofit	3% Nonprofit
19% Small Businesses	22% Small Businesses
11% Large Businesses	20% Large Businesses
4% Other	6% Other

SOURCE: Christopher Rinaldi, "STTR at DoD," presentation at the Academies STTR Workshop, May 1, 2015.

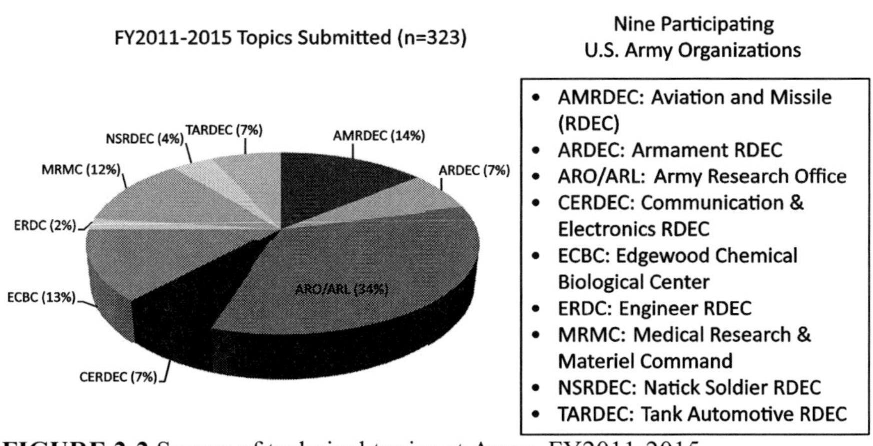

FIGURE 2-2 Source of technical topics at Army, FY2011-2015.
SOURCE: Bradley Guay, Army STTR Program Officer, "The Army STTR Program," June 2015.

[8]Discussion with Bradley Guay, Army STTR Program Manager, September 8, 2015.

Similarly, Navy STTR Program Manager Dusty Lang confirmed that STTR topics differ substantially from SBIR topics.[9] Although agency staff propose a wide range of topics every year, her office worked to ensure that the project involved an important role for RIs, such as access to cutting-edge research at research institutions. In contrast, the primary objective for SBIR was the immediate delivery of tools for the war fighter. She observed that STTR topics not only tend to focus on a lower technology readiness level (TRL), but also often provide applicants with more room for exploring still unformed or at least unconfirmed ideas.[10]

Department of Energy[11]

The Department of Energy manages the STTR and SBIR programs as identical administratively and functionally, which reduces administrative overhead. Other than the congressionally mandated differences between the programs, the agency does not see any significant or strategic distinctions between the two programs. Small businesses in both programs collaborate with RIs, and only a small percentage of STTR awards go to PIs employed primarily by an RI.

DoE offers two Phase I solicitations per year, both of which are open to applicants to either program. The solicitations propose the same topics for each program, and uniquely among the funding agencies, both Phase I and Phase II applicants can apply to either program or to both using a single application—as long as they meet the qualifications for both. Approximately as many applicants select both SBIR and STTR as do those who check just STTR.

Annual funding for DoE STTR is about $25 million. Awards are highly competitive, with a success rate of about 10 percent for STTR Phase I and 50 percent for STTR Phase II. DoE offers the same period of performance and award size for both programs: Phase I awards last 9 months for either $150,000 or $225,000. Phase II awards last 24 months, and may be for $1 million or $1.5 million.

Both the DoE SBIR and STTR programs have in recent years evolved to add emphasis on the commercialization of funded technologies. As a result, SBIR/STTR Phase I and Phase II applications must provide an initial evaluation of commercial potential, as the programs focus on providing seed capital for early-stage R&D that have commercial potential. Awards are, like those at other agencies, comparable in size to private-sector angel investments. However, both

---

[9]Discussion with Dusty Lang, Navy Program Manager, September 14, 2015.

[10]For an overview of the Technology Readiness Levels from basic research to operations, see https://cto.acqcenter.com/osd/portal.nsf/05A447FC781F1FA7852570C8006801E4/$FILE/trl_chart.pdf.

[11]Sources for this section are Manny Oliver, "The DoE STTR Program," presentation at the Academies STTR Workshop, May 1, 2015; discussions with DoE staff; and other material provided by DoE.

programs will deliberately accept greater risk than do angel investors, reflective of the agency focus on advanced energy technology.

Although STTR is designed to encourage collaborations between small companies and RIs, DoE notes that this is also the case for many SBIR projects as well. More than one-half of all Phase II SBIR projects include some funding for research institutions, and overall about 9 percent of SBIR funding goes to research institutions.[12]

The STTR program supports an extensive set of collaborations with DoE national laboratories, which on average constitute about one-third of DoE's STTR RIs, although their share has varied—from a high of 80 percent in FY1999 to a low of 13 percent in FY2014 (see Chapter 4 for a further discussion of National Labs and STTR).

Principal investigators for DoE projects mostly come from small business concerns, with only 13 percent coming from research institutions (including a small number primarily employed at National Labs)—even though STTR permits the principal investigator to be primarily employed at the research institutions. In 2014, only 3 out of 35 principal investigators in STTR were employed at the research institutions.

Like the other agencies, DoE has followed the 2011 reauthorization law to permit grantees to switch the type of program when entering Phase II. DoE has found that some STTR Phase I awardees are switching from STTR to SBIR during Phase II, but not the reverse. Since the program permitted such a switch in 2011, 10 out of 83 STTR Phase I applicants for Phase II funding have sought SBIR Phase II funding, and 4 have received it.

In 2013, DoE began a new technology transfer initiative, using the SBIR and STTR programs to transition technology developed at DoE National Labs and universities funded by DoE to the marketplace. The agency is prohibited by statute from using only the STTR program to foster technology transfer from its labs, so it uses both SBIR and STTR. This creates a range of new opportunities for collaboration between companies and both kinds of RIs, through the publication of specialized topics.

National Institutes of Health, Department of Health and Human Services[13]

The STTR program at the National Institutes of Health provides $95 million annually (compared with $691 million for the NIH SBIR program), an increase from $74 million in 2010.

NIH does not publish a separate solicitation for STTR with different

---

[12]Manny Oliver, private communication

[13]Sources for this section are Matthew Portnoy, "The NIH STTR Program," presentation at the Academies STTR Workshop, May 1, 2015; discussions with NIH staff; and other materials provided by NIH.

topics: the annual SBIR/STTR Omnibus Solicitation for grants[14] lists all NIH topics, and companies can apply for SBIR or STTR at their discretion (although not both, as is the case at DoE). NIH also publishes targeted funding opportunity announcement (FOAs) throughout the year, primarily from individual Institutes or Centers (ICs) seeking to focus their research on a specific technical area. Most of these funding opportunities are open to SBIR or STTR applications.

Overall, success rates for NIH STTR Phase I applications are slightly higher than for SBIR Phase I applications, but success rates have varied substantially by year (see Figure 2-3). The Fast Track program, which combines Phase I and Phase II applications, is available to both SBIR and STTR applicants and its success rates typically tracks those of Phase I.

In FY2014, STTR funded 160 out of 788 Phase I applications, and 37 out of 87 Phase II applications. Of the 60 Fast Track applications received, 5 were approved—a notably lower percentage than for SBIR. That year may have been an anomaly, because the success rate for FY 2013 was 28.6 percent—almost double that for SBIR.

In recent years, small business concerns and research institutions have received 50 percent and 45 percent on average, respectively, of STTR Phase I funding (see Table 2-3). For Phase II, small business concerns and research institutions have received about 58 percent and 40 percent, respectively, of total funding.

Universities accounted for about 79 percent of all research institutions funded through STTR Phase I in FY2014 (see Table 2-4). No federal labs or other Federally Funded Research and Development Corporations (FFRDCs, which are public-private partnerships that conduct research for the United States Government) have been funded since FY2010.

NIH-provided commercialization support programs opened to STTR awardees in FY2012, as provided for under the reauthorization legislation (P.L. 112-81). Since then, 22 percent of STTR Phase I awardees have participated in

TABLE 2-3 Small Business Concerns (SBC) and Research Institution (RI) Shares of STTR Funding, FY2010-2014

	Fiscal Year				
	2010	2011	2012	2013	2014
STTR Budget (Millions of Dollars)	74	73	85	81	97
Phase I: SBC Percent Share of Funding	55	56	53	56	54
Phase I: RI Percent Share of Funding	45	43	47	44	46
Phase II: SBC Percent Share of Funding	58	59	56	59	59
Phase II: RI Percent Share of Funding	40	38	41	41	41

SOURCE: M. Portnoy, "The STTR Program at NIH," presentation at the Academies STTR Workshop, May 1, 2015, slide 10.

[14]Although there is only one annual Omnibus Solicitation, there are three deadlines annually against which companies can apply for SBIR/STTR funding.

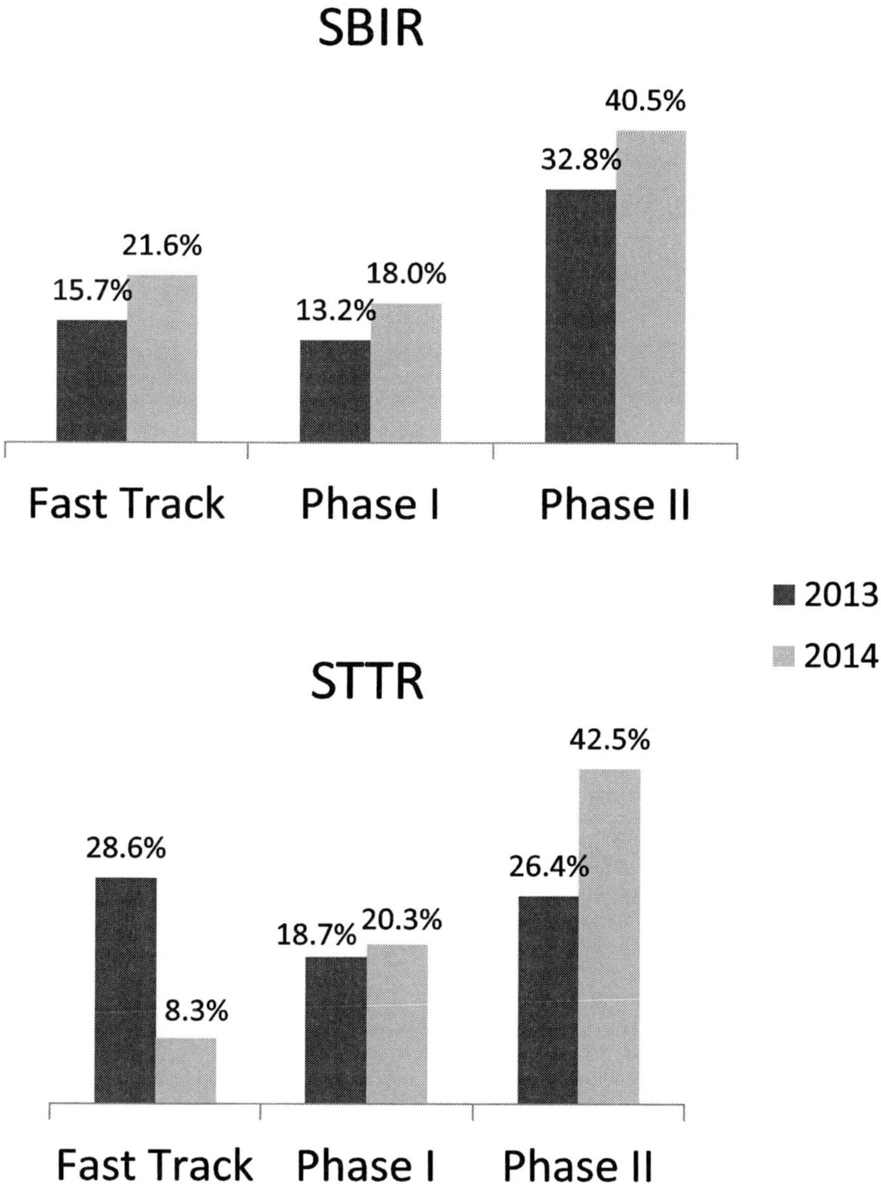

FIGURE 2-3 Success rates for STTR, SBIR, and Fast Track at NIH, FY2014.
SOURCE: Matthew Portnoy, "The NIH STTR Program," presentation at Academies STTR Workshop, May 1, 2015.

TABLE 2-4 Number of Universities and Other Research Institutions Funded by Phase through STTR, FY2010-2014

	Fiscal Year				
	2010	2011	2012	2013	2014
Phase I: Universities	118	81	123	114	130
Phase I: Other	24	17	22	30	34
Phase II: Universities	94	73	74	28	35
Phase II: Other	16	18	15	9	10

SOURCE: M. Portnoy, "The STTR Program at NIH," presentation at Academies STTR Workshop, May 1, 2015, slide 11.

the Phase I market assessment program operated on behalf of NIH by Foresight Inc., and 10 percent of STTR Phase II awardees have participated in the Phase II Commercialization Assistance Program operated for NIH by Larta Inc. In addition, two STTR companies sought funding for their own technical assistance.[15]

Two new initiatives may have a significant effect on STTR applications and companies: the NIH Center for Accelerated Innovation and the Research Evaluations and Commercialization Hub (REACH) award program. Both utilize the reauthorization of 2011 to establish "phase 0" proof-of-concept centers using, in part, STTR budget funding. Phase 0 programs provide support to innovation centers and hubs that develop promising technologies and position them for transfer into a start-up company or for licensing.

Since the programs were implemented about 18 months ago, NIH has made six grants to 14 partnering institutions belonging to consortia in Boston and in Ohio, and to five universities in the University of California system among others. All of the grants focused on the mission of the National Heart, Lung, and Blood Institute, which was recently joined in these programs by the National Institute for Drug Abuse. These programs are described in more detail in the 2015 Academies report on the NIH SBIR program.[16]

Overall, NIH staff have made it clear that, from the agency's perspective, there are no strategic differences between the SBIR and STTR programs, and to the maximum extent possible runs them in parallel (for example, offering only a combined solicitation). While they differ as outlined in the first section of this chapter, NIH does not use the programs for different purposes and operates them in close tandem, and specifically does not differentiate between the programs on the basis of their alignment with more or

[15]M. Portnoy, "The NIH STTR Program," presentation at the Academies STTR Workshop, May 1, 2015.

[16]For a more detailed description of these programs, see National Academies of Sciences, Engineering, and Medicine, *SBIR/STTR at the National Institutes of Health*, Washington, DC: The National Academies Press, 2015.

less commercial opportunities: they are equally expected to result in commercial outcomes.

National Science Foundation

Within the National Science Foundation, the SBIR and STTR programs are operationally located in the Directorate of Engineering's Division of Industrial Innovation and Partnerships (IIP) and are run as a single separate entity.[17]

NSF described its SBIR and STTR programs as "similar in almost every way," despite the statutory differences described in the overview section of this chapter. In addition, NSF awards SBIR grants of up to $150,000 for 6 months but STTR grants of up to $225,000 for 12 months. Additional funding for STTR and SBIR may be provided under NSF's Phase IB program, which matches funding provided by a third party with up to $30,000 more in NSF funds.

Phase II grants are identical for both programs—up to $750,000 for 24 months. Again, additional funding may be provided through NSF's Phase IIB program, which provides up to $500,000 in additional matching funds.

According to NSF staff, the agency experimented with targeted STTR topics that differed from SBIR topics in 2010-2012, but they were not popular with the business community (hence the low numbers of applicants; see Chapter 3). The agency has therefore returned to using the same topics for SBIR and STTR.

NSF describes STTR as a program for candidate companies whose technology is "earlier" on the TRL spectrum than those of SBIR companies. It sees STTR as closer to basic research than to development and earlier even than "early-stage" companies as seen from the perspective of a venture capital investor (see Figure 2-4). The NSF STTR and SBIR programs aim to move novel technologies to a point where private investors and industry can use them and move more rapidly toward commercialization. "Even an early-stage firm has a product and has revenue," said Barry Johnson, director of the Division of Industrial Innovation and Partnerships at NSF. "They just need help in scaling up. In the NSF STTR program we are trying to get our companies to that point."

The STTR program also works with other components and initiatives at NSF, including the iCorps program, which is described in the Academies 2015 report on the SBIR program at NSF.[18] STTR is also aligned with the Industry and University Cooperative Research Center program (I/UCRC), which has

---

[17]Sources for this section are Barry Johnson, "NSF STTR Program," presentation at the Academies STTR workshop, May 1, 2015; discussions with NSF staff; and other material from NSF and the NSF web site.

[18]National Academies of Sciences, Engineering, and Medicine, *SBIR at the National Science Foundation*, Washington, DC: The National Academies Press, 2015.

significant industry funding. The I/UCRC program addresses applied research, attempting to align what universities are doing with what industries are seeking—"university push and industry pull," according to Barry Johnson.

Discussions with NSF staff reveal that the agency places different emphasis on the programs, but considers them to be strategically similar and operates them together as closely as possible. The openness of the NSF SBIR program to early-stage ideas and to research proposals not tightly tied to specific solicitation topics suggests that the SBIR program at NSF is in some respects more closely similar in spirit to the STTR program than at other agencies.

National Aeronautics and Space Administration[19]

While NASA addresses the differing congressional objectives for SBIR and STTR, the agency is overall focused on generating technologies that can be adopted for use within the agency and directly serve the agency's mission.

NASA provides separate solicitations annually for SBIR and for STTR. The solicitations have different topics, and these are developed through somewhat different set of processes and procedures. In recent years Mission Directorates (MDs) have had increasing influence over SBIR, and SBIR topics have been approved by designated Mission Directorate staff, but the process for developing STTR topics is different.

For STTR, the Center Chief Technologist (CCT) at each of the NASA Research Centers plays a key role in developing topics related to the activities of their Center. These topics are usually aligned with the technology roadmaps and other strategic plans developed by the Centers and are primarily focused on the development of technology that will fit into the defined forward pathway. Unlike topic development for SBIR, the STTR approach does not seek technology that can be relatively quickly adapted for use within NASA. According to Joseph Grant, Deputy Program Executive, for the NASA SBIR/STTR program, "Each of the STTR topics and subtopics are mapped to the Space Technology Road Maps defined Technology Areas.[20]

The selection process is also strongly influenced by the Centers. The Technology Infusion Manager (TIM) at each Center works with the CCT to rank and prioritize proposals received in response to the solicitation, and it is the CCT who provides the final ranking and recommendations for funding to the SBIR/STTR Program Executive.

[19]Sources for this section include discussions with Robert Yang, NASA SBIR/STTR Program Executive, Joseph Grant, NASA SBIR/STTR Deputy Program Executive, other NASA staff, and materials provided by NASA or from the NASA SBIR/STTR web site.

[20]Joseph Grant, Deputy Program Executive, NASA SBIR/STTR program, presentation to the Academies STTR workshop, May 1, 2015.

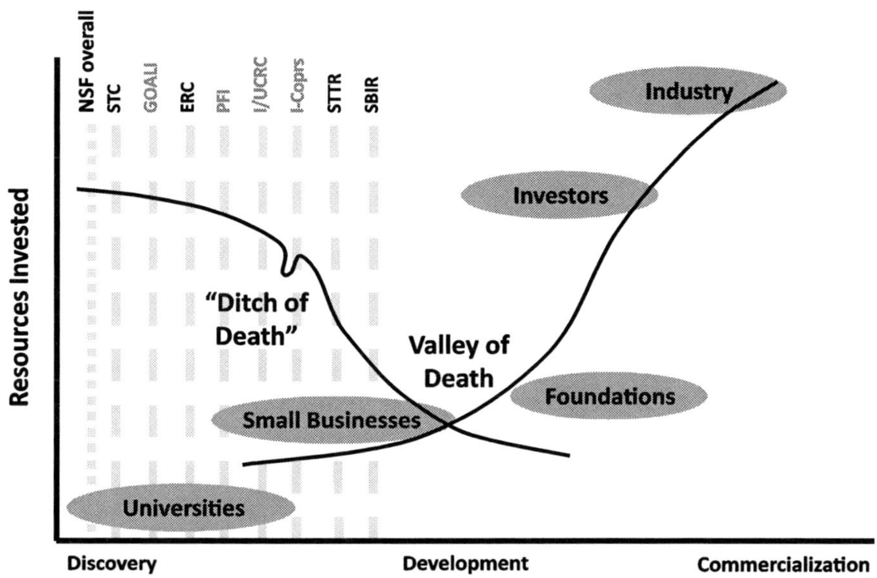

FIGURE 2-4 Strategic vision of STTR/SBIR innovation at NSF.
NOTE: Figure 2-4 shows STTR and SBIR in the context of other NSF innovation related programs, including the Science and Technology Centers (STC), Grant Opportunities for Academic Liaison with Industry (GOALI), Engineering Research Centers (ERC), Partnerships for Innovation (PFI), the Industry & University Cooperative Research Program (IUCRC), and the Innovation Corps (iCorps) program. STTR and SBIR are represented as being further along the development axis than these other programs.
SOURCE: Barry Johnson, presentation at the Academies STTR Workshop, May 1, 2015.

STTR phase I contracts at NASA are the same size as SBIR Phase I—$125,000—but can be performed over 12 months instead of 6. Phase II contracts are identical—a maximum of $750,000 over 2 months. Under reauthorization, companies can request to switch between SBIR and STTR after Phase I, but NASA has not yet received any such requests as of September 2015.

STTR projects are eligible to participate in agency programs designed to bridge the gap between technologies at the end of Phase II and the levels required for further commercialization. Figure 2-5 describes NASA's key programs, Phase II-E and Phase II-X.

While data on outcomes from the STTR program at NASA do not exist, the agency points to a number of successful projects that have made a positive impact on agency programs. For example, NASA STTR funding supported development of Techshot Inc.'s bone densitometer, which was the first x-ray machine on the space station. The technology allowed scientists to measure the bone density of rats under weightless conditions, following a $3.6 million Phase III award in 2012, for use on the space station in September 2014.

Phase II-Enhancement (II-E)

Phase II-E	Minimum non-SBIR/STTR Funding Required for Eligibility for Matching in Phase II-E	Corresponding SBIR/STTR Program Contribution	Anticipated Period of Additional Performance
	$25,000	$25,000	6-12 Months
	Maximum non-SBIR/STTR Funding to be Matched by SBIR/STTR Program in Phase II-E	Corresponding SBIR/STTR Program Contribution	Anticipated Period of Additional Performance
	$125,000	$125,000	6-12 Months

Phase II-eXpanded (II-X)

Phase II-X	Minimum Funding Required from non-SBIR/STTR NASA Source for Eligibility for Matching in Phase II-X	Corresponding SBIR/STTR Program Contribution	Anticipated Period of Additional Performance
	$75,000	$150,000	12-24 Months
	Maximum Funding Amount from non-SBIR/STTR NASA Source to be Matched in Phase II-X	Corresponding SBIR/STTR Program Contribution	Anticipated Period of Additional Performance
	$250,000	$500,000	12-24 Months

FIGURE 2-5 Phase II-E and Phase II-X programs at NASA.
SOURCE: Joseph Grant, presentation to the Academies STTR workshop, May 1, 2015. Access at http://sites.nationalacademies.org/PGA/step/PGA_160863.

Discussions with NASA staff indicate that they perceive a distinct strategic difference for STTR: it is focused on earlier stage projects, it creates a specific connection to RIs and it focuses more closely on the technology and less on the commercialization or NASA's use of the technology. It rebalances the program to at least some extent toward a more research-oriented perspective, and enables funding projects that have longer time lines and higher risk than SBIR.

IMPLEMENTATION OF STTR COMMERCIALIZATION BENCHMARKS

As with SBIR, reauthorization required that SBA and the agencies develop commercialization benchmarks for STTR. The SBA STTR policy directive includes two such benchmarks, which are identical to the language provided by SBA for the SBIR program, and which have now been implemented in the agencies' more recent solicitations:

(A) "The Phase II Transition Rate Benchmark sets the minimum required number of Phase II awards the applicant must have received for a given number of Phase I awards received during the specified period. This Transition Rate Benchmark applies only to Phase I applicants that have received more than 20 Phase I awards over the time period used by the agency for the benchmark determination.

(B) The agency Commercialization Rate Benchmark sets the minimum Phase III commercialization results that a Phase I applicant must have realized from its prior Phase II awards in order to be eligible to receive a new Phase I award from that agency. This benchmark requirement applies only to Phase I applicants that have received more than 15 Phase II awards over the time period used by the agency for the benchmark determination."[21]

Applicants must meet both tests. In reality, the bar is set very low: NIH uses 25 percent as the benchmark Phase I to Phase II transition rate ((A) above),[22] but the 2015 Academies study of the NIH SBIR program found that none of the top 20 companies in terms of number of Phase I awards won would come close to being excluded based on this benchmark rate.

At NIH, companies that must meet test (B) are required to show that they have received at least $100,000 in revenues or additional investment per Phase II award; thus a company receiving 15 Phase II SBIR/STTR over a 10-year period must show that it has generated a minimum of $1.5 million in commercialization.[23] These benchmarks are the same for all agencies.

That said, evidence from company case studies (see Chapter 4 on qualitative assessment) indicates that the new commercialization benchmarks are having a significant effect on company behavior, motivating them to focus more closely on the commercial opportunities that can flow from the SBIR/STTR programs.

It is worth noting that the Academies' methodology (described in Appendix A) goes substantially beyond the benchmarks identified by SBA, which should be regarded as a lower bound developed for administrative purposes. Outcomes for grants agencies (NIH, NSF, and DoE) will also inevitably differ substantially from those designed for contracting agencies who purchase the results of SBIR/STTR awards (DoD, NASA).

[21]SBA STTR Policy Directive, December 2014, Sections 4 (a) 3 (1) (A) and (B).

[22]NIH SBIR/STTR FAQs, https://sbir.nih.gov/faqs#reauthorization-sec21, accessed September 2, 2015.

[23]Even companies that are excluded under this rule are ineligible only for 1 year. See SBA STTR Policy Directive, December 2014.

3

Applications and Awards

This chapter describes and analyzes STTR awards and applications for the five study agencies—the Department of Defense (DoD), the Department of Energy (DoE), the National Aeronautics and Space Administration (NASA), the National Institutes of Health (NIH), and the National Science Foundation (NSF). Data for the STTR were drawn from the Small Business Administration (SBA) SBIR/STTR database for the program as a whole and from data provided by the study agencies for their own programs. With some variation for agency data, the analysis covers the period 2005 through 2014. This 10-year interval provides sufficient data to analyze trends and the evolution of the program. Additional agency-specific tables are included in Appendix F. A broader review of study data sources, methodological approaches, and potential biases can be found in Appendix A of this report.

This chapter reviews Phase I and Phase II STTR awards, in turn, and considers awards from a range of perspectives, including yearly trends, distribution by state, the impact of multiple awards to individual companies, and applications and success rates.

PHASE I STTR

Awards

In general, award and application patterns tend to follow those for SBIR.[1] Figure 3-1 shows total Phase I STTR awards for fiscal years (FY) 2005-2014 for all five study agencies.

---

[1]Unless otherwise stated, data on awards and amounts of funding for the entire program are drawn from the SBA SBIR/STTR awards database, available online at https://www.sbir.gov/sbirsearch/award/all. The SBA database houses the numbers reported by the individual agencies in a uniform format. SBA data do not always align perfectly with data provided by the agencies, either directly to the Academies or on their own web sites.

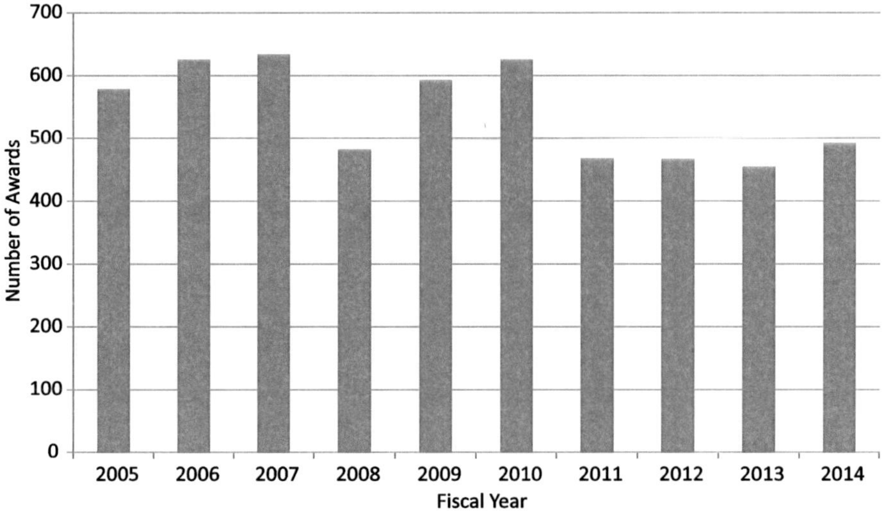

FIGURE 3-1 Phase I STTR awards, FY2005-2014.
SOURCE: SBA SBIR/STTR awards database.

Although overall numbers have remained approximately stable since 2008 (excluding the additional funding in 2009 and 2010 from the American Recovery and Reinvestment Act of 2009 [ARRA]), there has been considerable variation within agencies, especially DoD and NIH (see Figure 3-2).

Award numbers vary for many reasons and usually do not reflect deliberate agency decisions to increase or reduce the number of awards in a given year. Rather, variation can result *inter alia* from the complexity and riskiness of the research being funded, and the size and duration of awards. Variation can also result from changes in STTR program policy and agency budgets. For example, allowable award size has increased over time. Sequestration played a role in reducing award numbers in 2012, for example, while funding from ARRA increased award numbers at some agencies in 2009 and 2010. In some years, agencies might find that a preponderance of excellent Phase II proposals require funding, which may reduce the funds available for Phase I. In addition, shifts in application rates that are largely unrelated to the programs themselves might occur.

Awards by State

Award patterns by state are similar to those for SBIR, strongly correlated across and within states to reginal innovation centers.[2] Further

---

[2]National Research Council, *SBIR at the Department of Defense*, Washington, DC: The National Academies Press, 2014, pp. 30-32. Effective July 1, 2015, the institution is called the National

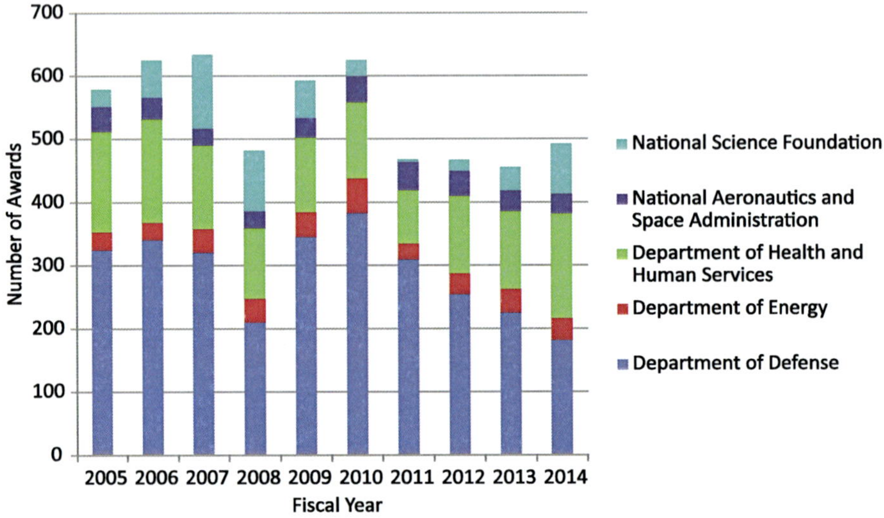

FIGURE 3-2 Phase I STTR awards by agency.
SOURCE: SBA SBIR/STTR database.[3]

differences in the rankings appear to reflect the distribution of partnering universities and research institutions (RIs)—not all states have strong RIs, and some highly respected research institutions do not participate heavily in STTR. For example, the Massachusetts Institute of Technology (MIT) maintains strong conflict of interest rules that make STTR less attractive for faculty than at some other research institutions.[4] Funding amounts, awards, and award shares for the top 20 states are shown in Table 3-1.

Collectively, the top 20 states accounted for 84 percent of all STTR awards during the study period. Conversely, the bottom 10 states (excluding Puerto Rico) accounted for 1.6 percent. The full table for all states is provided in Appendix F.

Awards by Company

The top 20 companies accounted for 11 percent of Phase I STTR awards during the study period (see Table 3-2).

Academies of Sciences, Engineering, and Medicine. References in this report to the National Research Council or NRC are used in an historic context identifying programs prior to July 1.
[3]In order to maximize comparability, data in Chapter 3 has been drawn primarily from the SBA SBIR/STTR database, which aggregates reports from all agencies. However, there are some discrepancies between data provided to the Academies by the individual agencies and the data gathered from SBA.
[4]Interview with Lita Nelson, Director, MIT Licensing Office, September 15, 2015.

TABLE 3-1 Phase I STTR Awards, Award Amounts, and Funding Shares for Top 20 States, FY2005-2014

State	Number of Awards	Amount of Awards (Millions of Dollars)	Share of Awards (Percent)
CA	910	124.0	16.8
MA	624	78.6	11.5
VA	355	42.2	6.5
TX	267	34.4	4.9
MD	249	35.3	4.6
NY	241	36.5	4.4
CO	220	27.3	4.1
OH	201	24.8	3.7
PA	184	28.4	3.4
IL	181	21.8	3.3
FL	154	18.7	2.8
NC	136	27.1	2.5
AL	134	15.4	2.5
MI	130	17.9	2.4
NJ	123	14.2	2.3
AZ	113	14.9	2.1
CT	90	15.2	1.7
GA	84	12.1	1.5
WA	83	12.5	1.5
OR	71	13.8	1.3

SOURCE: Small Business Administration, SBIR/STTR awards database.

WOMEN AND MINORITIES

Regarding participation of women and minorities in the STTR program, data from the SBA database are too inaccurate for conclusions to be drawn because comparisons with data supplied by agencies reveal significant discrepancies in this area. Awards for women and minorities will be discussed separately in Chapter 5.

PHASE II STTR

Awards

The number of Phase II awards made by each agency is to a considerable extent driven by the number of previous Phase I awards made. Overall, the number of Phase II STTR awards remained largely steady at

TABLE 3-2 Top 20 Phase I-Recipient STTR Companies, FY2005-2014

Company	Number of Awards
Physical Sciences	60
CFD Research	50
Luna Innovations	43
Intelligent Automation	43
Agiltron	35
Lynntech	33
Creare	33
Infoscitex	30
Charles River Analytics	30
Muons	25
Aptima	25
NextGen Aeronautics	24
NanoSonic	23
UES	19
Structured Materials Industries	19
Toyon Research	17
Aurora Flight Sciences	17
SA Photonics	14
Omega Optics	14
HYPRES	14
Top 20 Total	568
All Phase I STTR	5,420
Top 20 Percentage of Total Awards	10.6

SOURCE: Small Business Administration SBIR/STTR database.

slightly more than 200 per year, except for variations largely explained by the impacts of ARRA in 2009 and 2010 and sequestration in 2012 (see Figure 3-3).

Awards by State

Table 3-3 shows the total number and amounts of STTR awards made in the top 20 states. Collectively, these states accounted for 84 percent of all STTR awards made in FY2005-2014. As would be expected, the top 20 states for Phase II are closely similar (though not identical) to those for Phase I.[5]

[5]The order of states varies and Oregon (Phase II list) displaces Wisconsin (Phase I list).

BOX 3-1
Multiple STTR and SBIR Awards

There are a variety of pathways by which companies meet legislative objectives of the SBIR and STTR programs. As noted in Chapter 1 of this report, products do not develop according to a linear model, using one Phase I, one Phase II, and then perhaps a further contract to reach the market place. Some require that companies receive multiple awards. A cluster of awards may be necessary in some cases to develop core technology platforms and then devise different applications that arise from them.

Multiple awards are permitted by law and are often necessary to meet agency mission needs. For example, managers at DoD and NASA often turn to SBIR and STTR to solve their technical problems and develop new technologies for internal use. Both agencies publish long and detailed solicitations for technical help. Here, SBIR and STTR companies work within program rules when seeking out sequential opportunities to serve these agencies effectively. NSF, NIH and DOE similarly award multiple awards in selected cases to advance their missions.

Finally, SBIR awards are open to new entrants. About a third of awards go to companies who have not previously received an award from that agency. Multiple award winners thus do not crowd out new entrants.[a] For example in FY2013, more than one-third of companies offering SBIR/STTR proposals at NIH were new to the program, and about 23 percent of awards went to companies that had not received any prior awards from NIH.[b]

[a]National Research Council, *An Assessment of the SBIR Program*, Washington, DC: The National Academies Press, 2008. In addition, the report found that "The common perception about the prevalence of mills in the SBIR program—i.e., that they have captured a large percentage of the awards, that they rarely commercialize, and that they do not meet agency research needs—is not substantiated by the evidence." See p. 86.
[b]National Academies of Sciences, Engineering, and Medicine, *SBIR/STTR at the National Institutes of Health*, Washington, DC: The National Academies Press, 2015, p. 105.

Phase II STTR Awards by Company

The companies that are most successful at winning Phase I STTR awards are in general slightly more successful in winning Phase II STTR awards (see Table 3-4). Only two of the Phase I top 20 award winners are not among the Phase II top 20 award winners. For Phase II, the top 20 companies accounted for 264 awards (12 percent of total Phase II awards) over the 10-year study period, and, in comparison, for Phase I the top 20 companies accounted for 568 awards (11 percent of the total Phase I awards) over the 10-year period.

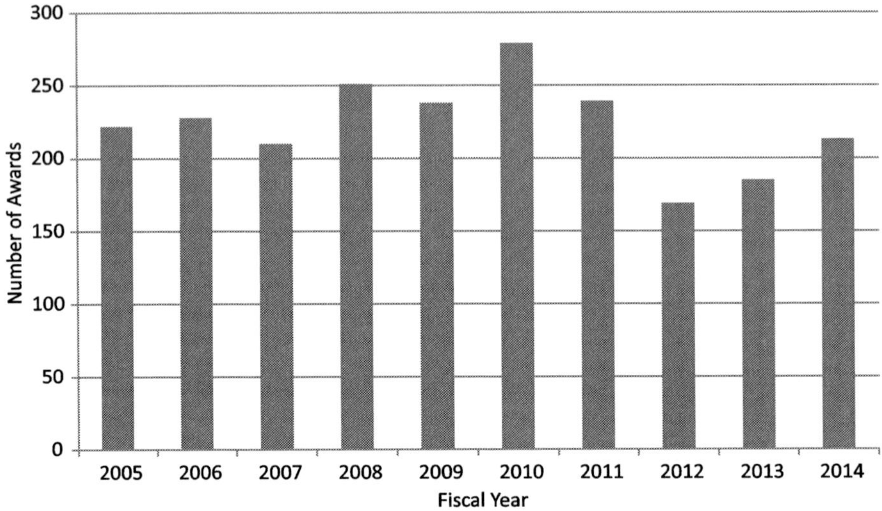

FIGURE 3-3 Phase II STTR awards, FY2005-2014.
SOURCE: SBA SBIR/STTR database.

TABLE 3-3 Phase II STTR Awards for Top 20 States, FY2005-2014

State	Number of Awards	Amount of Awards (Millions of Dollars)	Share of Awards (Percent)	Share of Funding (Percent)
CA	361	279.1	16.2	16.3
MA	265	195	11.9	11.4
VA	138	96.3	6.2	5.6
TX	113	82.6	5.1	4.8
NY	103	79.9	4.6	4.7
CO	97	69.3	4.3	4.0
OH	87	62.6	3.9	3.6
PA	87	69.9	3.9	4.1
MD	83	65	3.7	3.8
IL	79	57.6	3.5	3.4
FL	61	42.8	2.7	2.5
AL	56	44.6	2.5	2.6
MI	56	47.7	2.5	2.8
AZ	55	40.3	2.5	2.3
NJ	50	35.8	2.2	2.1
WA	47	37.8	2.1	2.2

State	Number of Awards	Amount of Awards (Millions of Dollars)	Share of Awards (Percent)	Share of Funding (Percent)
NC	43	37.2	1.9	2.2
GA	41	36	1.8	2.1
CT	32	26.5	1.4	1.5
WI	30	25.7	1.3	1.5
Top 20 Total	1,884	1,431.70	84.3	83.4
Total	2,234	1,715.90	100.0	100.0

SOURCE: Small Business Administration, SBIR/STTR database.

TABLE 3-4 Top 20 Phase II STTR Companies, FY2005-2014

Company	Number of Awards
Physical Sciences	28
CFD Research	25
Intelligent Automation	20
Luna Innovations	16
Aptima	15
Muons	13
Charles River Analytics	13
Lynntech	12
Creare	12
Agiltron	12
Aurora Flight Sciences	12
TDA Research	11
Toyon Research	11
UES	10
PRAXIS	10
Structured Materials Industries	9
NextGen Aeronautics	9
Infoscitex	9
Impact Technologies	9
Omega Optics	8
Top 20	264
All Phase II STTR	2,234
Top 20 Percentage of Total	11.8

SOURCE: Small Business Administration SBIR/STTR database.

4

Qualitative Assessment: Company and University Perspectives

This chapter addresses a range of program effects as described by executives of STTR companies and by agency and university technology transfer office staff. It also draws on textual responses by recipients to open-ended questions from the Academies Phase II recipient survey questionnaire (see Appendix A for detailed explanation of survey methodology).

This qualitative review provides context for the numbers discussed in the quantitative outcomes chapter and helps to explain why the program is so strongly supported by many award recipients.

This chapter is organized into three broad sections:

- Small business concern (SBC) perspectives
- Company concerns and recommendations
- STTR and the National Laboratories

Together, these sections provide the first wide-ranging publicly available feedback of the STTR program from program recipients and universities. Where appropriate, materials drawn from discussions with agency staff are also utilized.[1]

COMPANY PERSPECTIVES

STTR awardee companies provided input for this chapter in two ways: through case study interviews and through the Academies 2011-2014 Survey

[1]For this report, relevant staff were interviewed at NIH (including also interviews with a number of NIH institutes and centers); DoD (including interviews with component staff including Army and Navy); DoE, NASA, and NSF. This section also draws on material provided through the Academies STTR Workshop on May 1, 2015.

BOX 4-1
STTR Company (SBC) and Research Institution (RI) Consultations

Companies

Adelphi Technologies, Inc.; Calabazas Creek Research, Inc.; Creare, Inc.; Ekso Bionics, Inc.; Muons, Inc.; Nanosonic, Inc.; Physical Sciences, Inc.; Stratatech Corporation; Vista Clara, Inc.; Xemed, LLC; Xia, LLC.

A diverse set of case studies were selected for inclusion based on a range of selection criteria: extensive company experience with the program; strong comparative company experience with both SBIR and STTR; company experience with STTR (and SBIR) at a range of agencies; diverse company ownership; a range of company geographical locations—including location in well-known research clusters as well as in less concentrated areas of scientific expertise; and different company commercialization profiles. These profiles range from small research-oriented companies, to larger companies with strong track records as contract research organizations serving specific agency needs, to those focused tightly on the development and commercialization of specific products.

Company case studies are included in Appendix E of this volume. In all cases, appreciation is extended to the executives who took time to participate in interviews and provided further feedback through the review of preliminary drafts.

(discussed in detail in Appendix A). The survey provided opportunities for STTR recipients to describe differences between the SBIR and STTR programs, explain the impact of STTR on their companies, and suggest improvements to the program. A total of 305 textual comments were received, and 11 formal case studies were conducted. In addition, several companies that were earlier profiled for their experience with the SBIR program, in relation to the broader scope of the committee's study, were re-contacted and were provided the opportunity to also share their experiences with STTR. Case studies included companies with STTR awards from all five study agencies.

It is important to be cautious when contextualizing survey comments. Each company experience is somewhat unique, and as one recipient stated, "Every university partner is different." In addition, as discussed in Chapter 2 (Program Management), there are important differences in how the agencies run their programs. Indeed, Bradley Guay, STTR program manager for the Army, noted that even within the Department of Defense (DoD), the differences among

BOX 4-2
Different Kinds of Innovative Companies

Companies use STTR for a range of purposes, and agencies also use the mechanism for different reasons. As a result, some companies use SBIR/STTR to create a single product and then move on to commercialization, leaving the research stage behind. However, this is not an approach adopted by most companies, and in fact none of the companies reviewed in the case studies presented in Appendix E of this report followed this practice.

To some degree, companies respond directly to the needs of agencies, which are—for SBIR but not for STTR—explicitly defined as one of the Congressional objectives for the program. Especially at DoD and NASA, where the agency seeks to utilize the results of the research directly, STTR (as well as SBIR) is used in part as a specialized contracting tool to fund development of specific technical solutions. The companies who serve these needs often do so sequentially, proving a stream of innovations for agency use, but not necessarily commercializing them beyond sales to the agencies. Both programs are appropriately used to respond to agency needs. The companies may be commercially oriented, but often either require a cluster of awards to develop the core technology prior to commercialization, or continue to seek awards to develop new applications for platform technologies, or seek awards to develop a series of new technologies. Few companies remain sustainable on the basis of a single technology. Creating a sustainable technology basis is also an appropriate use of the program.

Finally, it is worth noting that company perspectives are in flux and the drive which exists across all the agencies to encourage commercialization has led companies to be more strategic in their use of the programs. Simply winning SBIR/STTR awards as a means of funding PhD researchers is increasingly seen as inappropriate both by the companies and by the agencies. As a result, even companies with a long track record of focusing on serving agency needs are increasingly using SBIR/STTR awards to bring new products to market.

components in operating their STTR programs are often greater than the differences between SBIR and STTR within a single component. He considered the STTR program at Army to be less similar to the STTR program at Navy than to the SBIR program at Army. Furthermore, because there are few STTR awards overall, it is difficult to make supportable generalizations about the program at individual agencies, or individual components within DoD.[2]

---

[2]Interview with Bradley Guay, Army STTR program manager, September 2, 2015.

SBIR and STTR: Differences from Company Perspectives

Some companies follow the Department of Energy (DoE), National Science Foundation (NSF), and National Institutes of Health (NIH) in seeing little difference between the programs. One survey respondent noted, "As a small business that mostly develops our own technology, the differences are only slight as we continue to collaborate with universities for needed services." There are similarities even in linkages to research institutions, the congressional objective for STTR. Many SBIR awards also involve research institutions as subcontractors, although their participation in SBIR awards varies substantially by agency.[3] Nevertheless, most respondents and case study interviewees did see differences between the programs.

Aside from the formal differences between the programs noted in Chapter 2, some survey respondents and case study interviewees said that, broadly speaking, where there were differences in topics, STTR topics tended to be more research-oriented and less focused on short-term commercialization. A respondent observed, "Topics with a high degree of fundamental research are appropriate to STTR. Topics with more focus on developing prototype hardware with nearer-term commercialization potential are better for SBIR due to the need for greater funding at the small business, and export control issues at universities." Similarly, another respondent observed, "The biggest difference lies in the topic definitions. The STTR projects tend to be much more academic, and harder to envision commercializing." Several program managers shared this perspective (see Chapter 2), although there is evidence that topic-level differences between SBIR and STTR are closing at some agencies. NSF experimented with separate topics for SBIR and STTR, but has since abandoned this approach, while at DoE the same solicitation is used for STTR and SBIR.

Survey respondents also observed that there were structural differences in the relationship between the small business and the research institution in the two programs. One noted, "In an SBIR award, the private business leads the effort and the University primarily supports that effort as a subcontractor. The STTR program is more of a partnership where the university personnel are given more autonomy and freedom. This gives the program a more academic flavor and the work is typically incorporated into one or more graduate dissertations."

Some respondents also saw a difference in the kind of technology sourcing involved. SBIR awards are used by small businesses to fund subcontracts at research institutions, which can perform specialized work or provide access to expensive equipment. In comparison, under STTR the research institution is sometimes itself the technology source. A survey respondent said,

---

[3]See National Research Council, *SBIR at the Department of Defense*, Washington, DC: The National Academies Press, 2014, and National Academies of Sciences, Engineering, and Medicine, *SBIR/STTR at the National Institutes of Health*, Washington, DC: The National Academies Press, 2015. Effective July 1, 2015, the institution is called the National Academies of Sciences, Engineering, and Medicine. References in this report to the National Research Council or NRC are used in an historic context identifying programs prior to July 1.

"In SBIR we can use the University as a subcontractor, taking advantage of specialized services that they may provide, or specialized expertise in a given area. In the STTR program, technology initially developed at the University is transferred to the Small Business for further R&D (in collaboration with the University)."

However, discussions with case study interviewees reveal that this generalization may be too broad and that technology can originate with the small business or the research institution under STTR. For example, Dr. R. Lawrence Ives (Calabasas Creek Research [CCR]) observed that his company had STTR awards for which the company was the primary technology provider and driver, as well as STTR awards for which the university researcher was the primary technologist. Similarly, Dr. David Green (Physical Sciences) said that the origin of the technology in his company's STTR projects might be the small business or the research institution.

Company Motivations

Although all STTR awards are made in response to a solicitation from the awarding agency, company motives for seeking STTR awards can be categorized into four primary areas:

- To access equipment and expertise for technology development to be found only at a research institution
- To help commercialize technology developed at the research institution
- To garner funding by addressing needs expressed through a solicitation
- To acquire indirect benefits from the program, such as reinforcing or developing links to an appropriate research institution partner, or identifying possible new employees.

Several companies with multiple STTR awards observed that they had at different times responded to each of these motives. However, several—such as Calabasas Creek Research, Vista Clara, and Xia—reported shifts over time, in particular away from more opportunistic motives and toward development of technology that was strategically important for the company. This shift may have resulted in part because of the new commercialization benchmarks introduced after the 2011 reauthorization. The companies mentioned above said that they now pay more attention to a topic's commercialization potential and are more reluctant to proceed unless it aligns with the company's overall commercialization strategy.

Access to University Technology, Expertise, and Equipment

Research institutions are repositories of technical expertise and in many cases equipment that companies especially at an early stage cannot afford to buy for themselves. As Dr. B. Lynn Allen-Hoffman (Stratatech) explained, the equipment in her laboratory at the University of Wisconsin represented an investment that was far beyond that available to a startup. Dr. Ives (CCR) noted that an STTR award allows a small company such as CCR to enter entirely new technology areas by tapping into expertise and equipment at a research institution. CCR's most recent STTR awards require equipment that it does not have and could not afford even with a Phase II STTR award but that is readily available at North Carolina State University.

Companies can tap equipment and expertise, however, through both SBIR and STTR awards. For a number of companies, the decision to seek an STTR award was based on the origin of the technology. As one survey respondent commented, "If the origin is from our business and we invite a university to join us, it is easier for us to frame the project as an SBIR. If it is the other way around, it would be easier to frame up an STTR project." Similarly, Dr. Charles Gary (Adelphi Technologies) noted that "the STTR program is primarily designed to allow [us] to directly acquire and commercialize technology developed at universities and national laboratories."

Enhanced Links Between Small Businesses and Research Institutions

Links to research institutions can benefit small businesses in many ways, while links of research institutions to small businesses offer a path to commercialization that may otherwise not be available to RI-sourced technologies. Dr. William Warburton (Xia) said that collaborations could be long running and very fruitful. For example, through an ongoing collaboration with Lawrence Berkeley National Laboratory, Xia was linked to a scientist aiming to move his technology out into the world. He provided access to instruments and considerable feedback. In exchange, Xia supplied him with next-generation equipment for his experiments. The collaboration has lasted 10 years.

More generally, STTR can help to strengthen linkages between the small business and the research institution. One survey respondent noted, "The STTR program is important to us because it helps gain and/or strengthen relationships with research universities in the area. This expands our technical horizons, and, most important, provides us with a glimpse and first bid on a pool of young, qualified new staff members, reducing our hiring timelines and risks." Another respondent wrote, "Interaction with research institutions provides access to knowledge, experience, and resources not available within the company."

Dr. Ives (CCR) explained that STTR provides an appropriate structure for partnering with research institutions and was particularly enthusiastic about the access it provides to the creativity and enthusiasm of graduate students. A survey respondent observed, "STTR relationships more easily support grad students. This is a long-term investment in highly qualified people with significant risk. It paid off for us, but there are no guarantees. The STTR mechanism is not for short term R&D but should be part of a long term strategy." Another respondent wrote, "STTR awards are beneficial to expand company relationships to potential future employees (graduate students) and outside perspectives."

This point can be aligned with quantitative data presented in Chapter 5—for example, Table 5-2 shows that 71 percent of respondents indicate that STTR has resulted in an enhanced relationship with the research institution.

Funding

Several of the companies interviewed indicated that STTR funding was critical for the company's formation and first steps toward commercialization. Dr. Bill Hersman (Xemed) said that the company was founded only when STTR funding was acquired. Dr. Kurt Amundsen (Ekso Bionics) said that STTR was especially helpful at the earliest stages of company development: Dr. Homayoon Kazerooni, a company principal, and other key staff were on the faculty at University of California, Berkeley, and STTR was the perfect bridge from academia to the company.

STTR funding can be used in other ways to advance the small business concern. For example, some companies reported that STTR validated their work and helped them to leverage additional investments from other sources. One survey respondent explained, "The funding provided by SBIR/STTR allowed my business to receive funding that could be used for credibility in leveraging for additional angel investment. Without the funding it would have been highly unlikely that this small business would have been created."

As companies mature and grow, STTRs can become less attractive. Dr. Amundson observed that Ekso Bionics now wants to move quickly and is less interested in academic research. It also does not wish to be so tightly coupled to university needs and interests.

Small Business Perspectives: Some Positive Conclusions

Although small businesses have many concerns about STTR, which are described in the following section, company interviewees and survey respondents also reported on STTR advantages.[4]

Companies reported that when partnerships with research institutions work well they can have a substantially positive impact on the company, providing access to unique technology and expertise, opening the door for the recruitment of future employees, and also providing a pathway for the commercialization of technologies being developed at research institutions. Moreover, the data reported in Chapter 5 reveal significant differences between

[4]See Table 5-2 on STTR impacts on SBC-RI relationships. Approximately 71 percent of survey respondents report an enhanced relationship.

SBIR and STTR regarding the role of the RI. These results reflect a key Congressional objective of the STTR program, which is to increase linkages between small businesses and research institutions. Table 4-1 shows responses to an Academies survey question about the type of linkages between the funded project and RIs, for both SBIR and STTR Phase II recipients.

COMPANY CONCERNS AND CHALLENGES

Projects that require teaming are likely to be more challenging for small businesses than are projects in which the small business has full control. Partnering with large organizations that often have different motivations and interests presents an additional set of hurdles. This section considers a number of challenges and concerns described by company interviewees and survey respondents.

Different Cultures: Small Businesses and Research Institutions

Several companies raised concerns about differences in perspective between small businesses and research institutions and, in some cases, differences in objective. Survey respondents observed that universities may not be well suited to partner with small companies. One survey respondent said, "Success of the STTR program depends a lot on the effort made by the University. One problem we consistently faced is that the academics tend to look at research as an open-ended problem whereas a small company has

TABLE 4-1 Research Institution (RI) Contributions to SBIR and STTR Projects

	Percentage of Responses	
	STTR Awardees	SBIR Awardees
The PI for this project was at the time of the project an RI faculty member	32.3	2.8
The PI for this project was at the time of the project an RI adjunct faculty member	2.3	5.3
Faculty member(s) or adjunct faculty member(s) worked on this project in a role other than PI	53.0	26.0
Graduate students worked on this project	51.1	20.3
The technology for this project was licensed from an RI	18.4	6.9
The technology for this project was originally developed at an RI by one of the participants in this project	29.3	11.1
An RI was a subcontractor on this project	70.3	25.8
None of the above	4.5	53.7
BASE: TOTAL RESPONDENTS ANSWERING QUESTION	266	1,795

SOURCE: 2011-2014 Survey, Question 71.

short-term goals and time constraints that may also not fit with the academic schedule."

The mandated participation of research institutions in STTR projects makes them significantly different from SBIR projects. The small business' mission is to create products and/or provide services to meet customer requirements; the RI's mission is to educate students (or deliver technology for the agency in the case of National Laboratories). In developing an STTR collaboration, the small business and research institution must work together to develop a statement of work that supports the small business' objectives while affirming the research institution's mission. This challenge does not exist for SBIR, which does not require a joint statement of work or a formal partnership with the research institution.

The research institutions perspectives on partnering with a small business can also vary. Dr. Ives (CCR) noted that CCR had differing experiences with research institutions. Some research institutions, such as North Carolina State University offered realistic licensing terms and welcomed collaboration with small companies. Others, however, did not appear to understand the limited resources of small businesses and demanded unrealistic upfront licensing fees and royalties.

Different Time Frames

Small businesses and research institutions can find themselves operating on different time frames within the same project. Small businesses are driven by the need to complete work on time and to reach commercialization as rapidly as possible. For them additional time means additional costs. Small businesses are also small and flexible and therefore can adjust to changing circumstances quickly. Research institutions in contrast are driven by the much slower calendar of annual budgets, academic research, and the needs of administrators.

These differences can present significant challenges to small businesses. One survey respondent explained, "Universities tend to work on a different (slower) time scale than SBIR subcontractors. There is no real urgency on the part of universities so work and deliverables are always slipping. Students working on STTR projects also do not place a high priority on the work." Another respondent wrote, "One must be very careful when choosing a partner for an STTR program. Many academic collaborators are not accustomed to working under tight deadlines or to supply results on a regular and timely basis."

Fixed Funding Shares and the Challenge of Enforcing Accountability

Some of companies surveyed noted that the structure of the STTR program contains a built-in rigidity: unlike SBIR, where small businesses are free to choose alternative subcontractors as needed, STTR requires that the same research institution remain a partner across all phases of the project. A survey respondent observed, "In the STTR award you benefit from technology developed at the research institution. However, it also entails a commitment by the Small Business to continue working with that research institution throughout the STTR project." As a result, if the research institution fails to deliver on its milestones, the small business remains responsible for the project and it must make good itself—which may be difficult or impossible—or must abandon the project. It is clear that the success of the STTR depends on the commitment of both partners and that the different perspectives of the partners can cause difficulties. Indeed, all partnerships work best when both parties are committed. While agencies do require written agreements between small businesses and research institutions, they cannot in advance determine the degree of commitment for a specific STTR project.

A survey respondent explained that he experienced projects in which one of the collaborators did not deliver, and the technical point of contact (TPOC) had no leverage against individual collaborators. Another respondent noted, "SBIR awards are much easier to manage, even with a university subcontractor. A non-performing SBIR subcontractor can be replaced or the work performed by the small business."

The STTR requirement that at least 30 percent of the funds flow to the research institution imposes significant fiscal drag on the project—at least from the small business perspective. A survey respondent wrote, "Getting the research institution to do what is expected on time and using their >30% of the budget is often difficult because of their higher-priority commitments. We often have to spend more than the amount we were allocated to get the work done in the time allowed, often with no funding extensions."

Other survey respondents said that the guaranteed funding reduces the RI's incentive to commit effectively to the project. A respondent wrote, "STTR mandates the minimum amount that must be funded to the university. This made [the RI] care less about producing results needed by the company. It also makes it almost impossible for the company to operate profitably, with more than 30% of revenue given to [the RI] for results that may or may not meet company needs." Another respondent explained, "It is hard enough getting a technology developed, prototyped, and delivered to a user who cares on limited funds. Having to fund a University makes it extremely difficult to achieve a positive outcome."

Other survey respondents noted that the mandated division of funding may not be optimally designed for the project: "Under STTR, the small business has less flexibility to determine the optimum split of funding between the business and the research institution. In some cases, this limitation is not an issue, but in other cases it seriously hampers the small business' ability to attain the highest level of technology readiness at the conclusion of the project." Although some companies (such as the one that commented above) complained that research institutions receive funding that could be better spent elsewhere, other companies expressed a wish to provide the research institution with more funding than the maximum allowed, for example, a life sciences company needing to engage a contract research organization with expertise in running clinical trials.

Intellectual Property (IP) and Licensing

STTR awards require that the small business and the research institution reach an agreement on intellectual property (IP). In the case of National Laboratories, these agreements are included in the cooperative research and development agreement (CRADA) and are effectively standardized. Universities, however, can adopt a variety of approaches and policies.

Discussions with both companies and technology transfer offices (TTOs) revealed a wide variation in attitudes and polices at research institutions toward licensing their intellectual property and the IP agreement that is mandatory for STTR awards. One respondent explained, "For an STTR, negotiating the appropriate intellectual property agreement can be problematic. Typically, the intellectual property departments of academic and non-profit research institutes have very little experience in dealing with STTRs. Developing the agreement can be a long process."[5]

Dr. Ives (CCR) observed that publishing is the primary objective of universities and graduate students. It is therefore important to understand this need and to act accordingly by providing opportunities to publish without compromising company IP. He believes that it is entirely possible to balance these needs, as evidenced by his company's record of publications related to university collaborations.

A survey respondent was more critical, noting that "Working with Universities on IP issues is typically a nightmare. We now only use Universities for their testing facilities.... University involvement for most of our commercial customers is typically a poison pill and so we only use universities when working on Federally-funded programs." Another respondent wrote, "Negotiating licensing with the research institution can sometimes be difficult as the technology is usually at the moment worthless and very hard to know what the future market will be."

[5]At present, agencies are not set up to evaluate the extent of the 'STTR-friendliness' of a technology transfer office, and there are only a very limited number of data points for any individual RI/agency pairing. This means that agencies cannot as yet 'reward' technology transfer offices with STTR friendly policies.

Discussions with university technology transfer offices indicated that different universities saw their missions in this area quite differently. Land-grant universities such as North Carolina State saw their mission as explicitly involving economic development within the state. As a result, they were focused more on finding ways to utilize additional funding from SBIR and especially STTR, while still addressing conflict of interest concerns and other issues related to IP. Other universities did not see economic development as a mandate, and were much more cautious about potential conflicts of interest. They saw no structural role for STTR and SBIR, and were deeply concerned about potential problems, for example, faculty employing graduate students in the private sector.

Beyond intellectual property agreements with the research institutions, patenting in general is a challenge for small companies. As Dr. Ives (CCR) noted, his company has historically used patents to protect its IP (see the list of CCR patents in the CCR case study provided in Appendix E). However, he is concerned that the rising costs of patents means that in the future CCR will have to be much more selective about which technologies it seeks to patent. He is therefore a strong supporter of a recent DoE initiative to permit a limited amount of Phase II funding to be used for patent-related costs.

Contracting Concerns

In general, small businesses view contracting as significantly more difficult for STTR than for SBIR. This is especially true when the research institution is a National Laboratory, because the multiple layers of management and the need for signoff by both the laboratory management and the agency can delay matters significantly. Such an arrangement also requires a cooperative research and development agreement (CRADA)—see "STTRs and the National Laboratories," below.

A survey respondent noted that university participation inevitably complicated the contracting process. Another respondent observed that "the collaboration element is harder to manage… executing a timely subcontract can be difficult as some research institution contracting departments are slow." Still another noted that "once an research institution is involved, the degree of difficulty for submitting a proposal and managing an award go up significantly."

Universities also have their own issues and concerns related to contracts administration. A survey respondent observed that "SBIR is always easier where as [sic] with an STTR I have to manage the conflict of interest requirements of the University."

The administrative burden imposed by STTR awards can be challenging: "Without the help of our experienced principal investigator and to a lesser extent the Project Manager, the rules and reporting requirements can be quite formidable for a small company like ours. I know there are rules and regulations behind why things are done, but simplifying the interface between NIH and the recipient would be helpful." Forty-five percent of survey respondents indicated that SBIR was easier to manage than STTR, while only 3 percent thought the reverse was true.[6]

Administrative issues do not arise only in relation to research institutions. Dr. Hersman (Xemed) explained that for one of his company's projects the DoE contracting officer did not understand the program: "Xemed had received an STTR award at DOE. Then a purchasing agent within DoE unfamiliar with STTR thought STTR could not have an academic PI. The agent halted the grant, and this could not be reversed even though DOE SBIR/STTR program officers agreed that the decision was wrong. I had to mortgage my house, and this almost sunk the company."

Dr. Hersman also discussed ambiguity in the extent to which the PI can be located at an academic institution provided there is a "special relationship" with the company. In the past, Xemed solved this potential problem by appointing the PI to a position on the company's advisory board.

The New Commercialization Metrics

It became evident from discussions with company executives that the new commercialization metrics instituted by the agencies after the 2011 reauthorization have had a distinct effect on SBIR/STTR companies. These impose new metrics related to the Phase I/Phase II transition metrics and also mandate that recipients of more than 15 Phase II awards over 10 years show evidence of at least $100,000 in commercialization per award). While these requirements are not especially burdensome, and impact few companies, companies in general clearly feel under pressure to show more commercial activity and results.

Dr. Gary (Adelphi Technology) explained that in recent years the company has completed its evolution from a research-oriented company into a more product-focused company and has redirected its focus on the development and then sale of compact neutron generators. Dr. Ives (CCR) said that the company has recently become more selective about the topics that it pursues, focusing more on commercial opportunities.

The new benchmarks have had an adverse impact at some companies. Dr. Roland Johnson (Muons) said that the new metrics "almost killed us." Immediately after program reauthorization, Muons' proposals to DoE were marked as noncommercial and rejected. The company developed more commercial activities, including contracts with Fermilab, Toshiba, and Niowave, and is now, after receiving $3 million in contracts over the past 5 years, again eligible for DoE awards.

Dr. Johnson (Muons) noted that the company changed its approach even for successful projects: its GBeamline software was given away and is in use at almost all the top high-energy physics laboratories worldwide. Muons will, however, license its next product on a more commercial basis, and it plans

---

[6]2011-2014 Survey, Question 80.

to spin off a new business focused on software support. Muons is actively seeking funding for a major initiative in partnership with large companies focused on nuclear reactors for the Navy.

Challenge of Indirect Costs

A less obvious challenge in managing an STTR award relates to the requirement that at least 30 percent of project funds must go to the RI, and that typically about 30 percent of this funding is used by the research institution to cover indirect costs.[7] That amount is therefore not available for the project budget. As one respondent noted, "STTR awards are less valuable to the business because so much of the award is taken by indirect costs at the research institution." As on average research institutions receive about 40 percent of an STTR award, then the indirect cost or "tax" on STTR projects for research institutions is approximately 12 percent of overall funding: a significant slice of the pie, especially for projects with tight budgets to begin with.

Issues similar to these are being addressed elsewhere. For example, the European Union has instituted a firm flat rate of 25 percent for its 2020 Horizons research program, and Japan's government programs pay 30 percent.[8] Some survey respondents indicated that they would support imposition of such caps on indirect costs for STTR. A respondent describes the essence of the issue at a particular agency: "NIH should allow us to exclude indirect costs from the budgets of research institutions (as the agency does for RO1 basic research awards). Indirect costs at most research institutions exceed 50% at this point and they take a significant bite out of the research institution side of the budget. Another alternative would be that taken by NIH to limit indirect costs to ~20% of the research institution budget for SBIR/STTR applications."

Limits of Commercialization Under STTR

Some survey respondents suggested that efforts to focus STTR more closely on commercialization are misplaced. They noted that universities are often poorly positioned to support commercialization and present barriers to its successful pursuit. A survey respondent suggested that the program should focus on other benefits of small business-STTR interaction: "Typically Universities are way too far away from being ready to commercialize a technology for these questions to be valid. The questions should have focused around 'did the STTR provide the opportunity for the Small business to: Gain experience in a new technology area? Work with graduate students who later became employees? Access assets that a small company could not typically use in conducting R&D?' It is here that the value of the STTR program lies."

---

[7]Heidi Ledford, "Indirect costs: Keeping the lights on," *Nature*, November 19, 2014.
[8]Ledford, "Indirect costs."

STTR can have profound impacts that are not measured in standard commercialization metrics. For example, Dr. Warburton (Xia) explained that the company has sold approximately $20 million in instruments for synchrotrons. The latter cost $500 million to build and approximately $200 million per year thereafter, and a large percentage of the research undertaken with these systems required Xia instruments. Fluorescence experiments would not run at all without them, and overall productivity would be maybe 10 percent of what it is today. Similarly, Xia developed instruments for measuring background radiation that have been used for validating compliance with nuclear test-ban treaties—another market with minimal sales but potentially large social impacts.

Are Research Institutions Value for the Money?

Some survey respondents clearly preferred SBIR because the company itself is more efficient in conducting commercially oriented research. Dr. David Walsh (Vista Clara) explained that while he does not object to partnering with academic institutions on occasion, in most cases Vista Clara would have done a better job without them. According to him, in only one out of the seven to eight partnerships formed for SBIR/STTR did the university add real value.

A survey respondent echoed his point: "Rather, in about 3/4 of the cases, my company would have had better value if the topic was an SBIR topic and award, rather than an STTR award. This is primarily because company staff could have performed (all or a portion of) the research more efficiently than the university."

Case study interviewees were asked whether, if they had a choice, they would prefer to use SBIR or STTR as a vehicle for research. There was a strong preference—especially among researchers with considerable STTR experience—that they would prefer to retain more control of the project by using SBIR.

UNIVERSITY PERSPECTIVES

Clearly, STTR can be very beneficial to academics seeking to transition their technologies from the laboratory into a commercial company. Ekso Bionics found that early STTR awards allowed key staff to transition from the university to the company while principals retained their positions on the faculty at University of California, Berkeley. Dr. Allen-Hoffman stated that she was able to act as PI on some key research projects funded by STTR while retaining her post at the University of Wisconsin.

However, the perspectives of university technology transfer offices are not as clear. Although the focus of this study has been on the small business concerns and the agencies that provide STTR funding, a preliminary set of discussions were held with selected technology transfer offices. Perspectives on STTR (and SBIR) vary widely. Some technology transfer offices focus on what might be called Phase 0, that is, preparing academics to submit successful

SBIR/STTR awards or finding alternative funding for very early-stage development. Other technology transfer offices—such as that at the University of Washington—were developing a full set of supports that connected spinoffs from the university to SBIR and STTR across the entire range of program activities and beyond. Other universities had different mechanisms for providing that type of support—for example, the Wisconsin Alumni Research Fund (WARF), in parallel with STTR and SBIR, supported Stratatech's development of path-breaking treatments for burns.

In some cases, however, technology transfer offices were much less engaged with STTR than with other issues, which stemmed in part from strong university conflict of interest rules. For example, faculty of the Massachusetts Institute of Technology (MIT) are not permitted to use university laboratories to support any research needed by a professor's for-profit company, and the professor must remain a full-time faculty member. As a result, according to Lita Nelson (director of MIT's technology transfer office), STTR and SBIR play a very small role in commercialization of MIT technologies.[9]

The issue of conflict of interest has to be addressed at every university, as do university policies on royalties and other payments for IP. As Dr. Ives (CCR) observed, some universities understand how to conduct business with small businesses, and others do not. Demand for high levels of payment up front, immediate reimbursement of patenting expenses, and uncertainty about royalty payments can make universities less attractive partners for small businesses. In contrast, the University of Minnesota has developed a very clear roadmap defining how small businesses can utilize university IP, and as a result the number of agreements with industry partners has grown rapidly since the new program was announced.[10]

STTR AND THE NATIONAL LABORATORIES

At DoE, about one-third of STTR awards involve partnerships with National Laboratories. This is not the case for other agencies, although all have STTR awards for which a National Laboratories is the research institution partner.

Linking small businesses and the National Laboratories involves substantial structural difficulties. The latter are usually operated by government contractors—nonprofits such as Battelle—rather than directly by government staff. For the National Laboratories, even a Phase II STTR award is a small amount of money. Thus, while scientific staff may be enthusiastic about working with an small business on an exciting project, administrators may see only the burden rather than the opportunity. Administrative costs for the

[9]Lita Nelson, director, MIT Technology Licensing Office, interview on September 15, 2015.
[10]Jay Schankler, director, University of Minnesota Office of Technology Commercialization, interview on September 15, 2015.

laboratory can effectively swallow all of the funding that might be provided to the laboratory under a Phase I award.

More generally, there are few incentives for National Laboratories to collaborate with small businesses. Dr. Warburton (Xia) said, in the best of cases, the laboratory scientists see STTR as a means of supporting their research program, in exchange for providing the company with technical support. In other cases, laboratory staff see the program as a means to generate funds, and often have no interest in commercial outcomes or even their partner's interests.

In addition, laboratory procedures are cumbersome. All teaming agreements require a CRADA, and in the case of SBIR-STTR, each phase requires a separate CRADA. Furthermore, although the basic structure of the CRADA almost always follows the standard Stevenson-Wydler model contract, according to Dr. Johnson (Muons), any change to the statement of work must be approved not only by the laboratory staff but also by the DoE cognizant officer who controls laboratory activities on behalf of DoE.

Cumbersome procedures can lead to substantial delays. In fact, as Dr. Johnson pointed out, CRADA approvals can take months. As a result, small companies working with National Laboratories must develop mechanisms for managing substantial volatility in funding flows, which could be disastrous. He also noted that delays by the laboratory in approving a change to the statement of work could result in the laboratory and the small business working on different timelines, and therefore the laboratory being a year behind the agreed timeline.

Working with the National Laboratories presents other challenges as well. Several interviewees explained that, as laboratories are fundamentally research organizations, they work on principles oriented around the free exchange of information and ideas, eventually leading to peer-reviewed publication of scientific and technical advances. The small business may, however, need to maintain closer control of IP developed under an STTR award, either through patents or trade secrets, and this need to control the flow of information can create significant cultural tensions with normal laboratory operations. Dr. Warburton (Xia) noted that each laboratory has its own culture; Xia worked quite successfully with Pacific Northwest National Laboratory and Lawrence Livermore National Laboratory, but not with other laboratories.

Several survey respondents and interviewees noted that STTR agreements with National Laboratories were less enforceable than SBIR subcontracts. Under SBIR, the small business concern can simply refuse to pay or switch to another supplier if the laboratory fails to deliver the technology or work.

Under STTR, the small business concern is committed to the research institution for the entire Phase I/Phase II cycle and has no recourse if the research institution fails to deliver. As Dr. Johnson (Muons) noted, in such circumstances, the small business would have to do the work itself—it could not fire or sanction the research institution. Dr. Warburton (Xia) said that his company's collaboration with Brookhaven National Laboratory was especially poor, with no accountability for the project at the laboratory. The Laboratory's role was to develop a specific mechanism, but it did not deliver.

Still, interviewees and survey respondents provided cases of highly successful STTR partnerships with National Laboratories. These seemed especially likely to succeed if the small business had a deep understanding of the laboratory. In several cases—such as Muons—at least one small business executive had worked for many years within the National Laboratory in question and therefore was highly knowledgeable about laboratory culture and procedures.

Finally, commercialization is not the only positive outcome from STTR. When working with the National Laboratories, some companies view their mission as largely outside of commercialization. Dr. Johnson (Muons) said that his company focuses on serving the technical needs of DoE and in particular the laboratories, much as some SBIR companies serve DoD. He believed, based in part on his extensive experience as a laboratory employee, that a small firm could provide creative solutions that were difficult or impossible inside the laboratories. For example, Dr. Ives (CCR) said that his company partnered with the SLAC National Accelerator Laboratory to improve the performance of cavity resonators used in linear accelerators. Stronger electric fields within the resonators means accelerators can be shorter, potentially saving millions of dollars in construction costs. However, these cost savings did not show up in the commercialization data.

Although this report is not an assessment of the National Laboratories, it is worth noting that some survey respondents see significant changes in the laboratories' attitudes toward STTR. Dr. Johnson (Muons), for example, observed that the laboratories had traditionally seen STTR (and SBIR) as a tax on research funding, but this perspective has changed in recent years. His view is that the laboratories have become more interested in finding ways to use STTR (and SBIR) awards to meet their technical needs.

STTR IN PERSPECTIVE

The STTR program was created in 1992 to improve the participation of research institutions in small business innovation. Making the case for STTR, Jon Baron noted at the time that "the purpose of the program is to create an effective, systematic vehicle for moving commercially promising ideas from the nation's research institutions to the marketplace."[11]

This review of the STTR program is the first to assess the performance of this vehicle since the passage of the legislation. This study finds that while STTR does make the grade in addressing its legislative objectives, its potential to transfer technology is not being fully realized because STTR is hard to use. As information from the case studies, survey responses, and agency interviews

[11]Jonathan Baron, "The Small Business Technology Transfer (STTR) program: Converting research into economic strength," *Economic Development Review*, 1.4 (Fall 1993):63.

presented in this report show, small business entities often find that partnership requirements for STTR are onerous and agencies find STTR cumbersome to administer.

Small Business Perspectives

The case studies and survey of small businesses underscore contrasting perspectives on the value of the STTR program. Some note that STTR can act as a bridge between research institutions and small business concerns: For example, Dr. Hoffman of Stratatech points out that STTR performs an important bridging function. She adds that this bridging function was not available through SBIR because of the 51 percent requirement for the principal investigator to work at the small business. Other case studies show that STTR strongly encourages a range of linkages between the small business and the research institution. For example, Dr. Ives of Calabazas Creek Research notes that bringing graduate students in at an early stage benefits small businesses in several respects, as they bring new energy and ideas and in some cases could be recruited to become employees at a later stage.

On the other hand, the case studies and survey responses show that, overall, STTR projects are harder for small businesses to manage than SBIR projects. In particular, the mandatory connection with a research institution, required for STTR—and especially the required intellectual property agreement—is a significant challenge for small businesses, where the success of this collaboration depends to a large extent on the commitment of the research institution to the project. This can be especially problematic where the research institution is not familiar with or particularly interested in the STTR mechanism.

Many small businesses find SBIR easier to use than STTR. Referencing 2011-2014 Survey question 80 (See Appendix C), 45 percent of the 211 survey respondents who had sought and received both SBIR and STTR awards, found that STTR was harder to manage than SBIR. Only 2.8 percent found STTR easier to manage than SBIR.

Typically, STTR technologies are at a very early stage of development when it is not clear how commercially successful they will be or what the size of the relevant market will be. Yet some universities require significant royalty payments upfront, as well as payment for patenting costs and other expenses. Small businesses find that this approach to licensing makes it difficult for them to work with universities, making a successful partnership through STTR less likely.

The case studies also illustrate how STTR can be a useful mechanism through which scientists and engineers at National Laboratories can pursue exciting but less advanced ideas. However, small businesses also expressed frustration with the difficulties of holding National Laboratories (as well as other research institutions) accountable for STTR deliverables. Small businesses also highlighted the cultural differences between the more open culture of some universities and the relatively closed commercial culture private businesses.

Agency Perspectives

The contracting agencies (DoD and NASA) use STTR and SBIR for strategically different purposes than the other three agencies that operate STTR. Program managers at NASA and at DoD (in particular the Army and Navy) see STTR as filling a gap between basic research and acquisition programs. STTR is now used by the Navy and Army as a means of addressing potentially valuable technologies at lower technology readiness levels (TRLs) that are not necessarily aligned immediately with the needs of the acquisition programs. In effect, STTR is used at DoD to undertake preliminary work on technologies that may eventually become extremely valuable. In contrast, the SBIR program is increasingly aligned with the immediate needs of the acquisition programs at DoD.

NASA similarly uses different mechanisms to develop topics for STTR, and strategically focuses STTR on lower technology readiness levels than SBIR. As a result, NASA and DoD program managers view the STTR program as a significant conduit between their agency and leading research universities. Program managers tell us that the STTR program provides a mechanism through which these procurement agencies can explore cutting-edge technologies in a research environment.

In contrast, granting agencies (NSF, NIH, and DoE) see little strategic difference between STTR and SBIR. Their SBIR and STTR programs do not, in practice, focus on different objectives, and to the maximum extent possible managers at these three agencies operate their SBIR and STTR programs in parallel or even a combined entity where feasible. As a result, there is minimal or even zero operational differences in the management of the STTR and SBIR programs at the granting agencies. There are no separate sets of topics, the application process is identical, and award management is essentially identical. DoE goes further in allowing companies to apply simultaneously for SBIR and STTR.

SBIR and University Links

When STTR was established, it was believed that SBIR did not address the need to transfer technologies from universities and other research institutions. Jon Baron argued that "Whereas SBIR exploits commercially promising ideas which originate in the small business community, STTR will use the successful SBIR approach to exploit a vast new reservoir of commercially promising ideas which originate in universities, federal laboratories, and nonprofit research institutions."[12] At that time, there was no data on the extent of SBIR's role in reaching out to research institutions.

[12]Ibid.

Recent assessments by the Academies show that SBIR actively connects innovative small businesses to research institutions. The 2008 summary report on the SBIR program found that 36 percent of survey respondents across the five major agencies reported involvement with a research institution in their SBIR project.[13] The Academies 2015 report on SBIR/STTR at NIH, reports that 63 percent of SBIR Phase II respondents participating in the NIH SBIR program indicated a university connection.[14] And the Academies 2015 report on SBIR at NSF, finds that 58 percent of SBIR Phase II respondents participating in the NSF SBIR program indicated a university connection.[15]

This strong level participation by research institutions in some SBIR programs comes despite the fact that the legislation governing the SBIR program requires that the principal investigator be employed at least 51 percent time at the small business concern. This can cause significant conflict for faculty whose contracts with their institution often require that they be employed at least 51 percent time at the university or laboratory. STTR permits a lower level of employment but, as noted above, poses other obstacles for participation.

To improve the transfer of technologies from research institutions to the marketplace, existing impediments for collaboration between research institutions and small business concerns must be tackled. As discussed in Chapter 6, this would include reducing difficulties in developing and maintaining contractual agreements within STTR. Changes to SBIR, such as selective waivers to the SBIR 51 percent small business employment requirement, may also be considered.

---

[13]National Research Council, *An Assessment of the SBIR Program*. Washington, DC: The National Academies Press, 2008, p. 245.

[14]National Academies of Sciences, Engineering, and Medicine, *SBIR/STTR at the National Institutes of Health*, Washington, DC: The National Academies Press, 2015, p.124 of the pre-publication report.

[15]National Academies of Sciences, Engineering and Medicine, *SBIR at the National Science Foundation*, Washington, DC: The National Academies Press, 2015, p. 100.

5

Quantitative Outcomes

While there are many success stories from individual projects and companies published on agency websites, quantitative data on outcomes is difficult to acquire. Yet the need to move beyond individual cases remains; this chapter presents a more systematic view of program outcomes through the use of survey tools.

Tracking SBIR/STTR outcomes is a work in progress. Some agencies have their own outcomes collection systems and metrics, while others are seeking to connect to the emerging data collection system at the Small Business Administration (SBA). However, the SBA data collection system is not yet complete, and no outcomes data are available from SBA as of August 2015. Thus, analysis of outcomes presented in this chapter is based on the 2011-2014 Survey of Phase II SBIR/STTR recipients at the five major research agencies (Department of Defense [DoD], National Science Foundation [NSF], National Institutes of Health [NIH], Department of Energy [DoE], and National Aeronautics and Space Administration [NASA]). The survey covered awards made in fiscal years (FY) 1998-2007 for DoD, NSF, and NASA and FY2001-2010 for NIH and DoE, inclusive. These agencies have all operated the STTR program since 1993.

The survey focuses primarily on outcomes related to the congressional objective for the program, as well as on commercialization outcomes and longer term impacts on the recipient companies themselves. Appendix G includes more detailed tables and analysis. Where useful, results for SBIR awards are also provided for context.

A detailed description of the methodology underlying the Academies survey is provided in Appendix A of this report. The full text of the survey questionnaire is provided in Appendix C. Questions in general focused on the individual project funded under the program. For the case of information gathered about companies as opposed to individual projects, multiple responses from the same company were aggregated and then averaged to provide a better

BOX 5-1
Survey Response Rate and Non-Respondent Bias

As noted in the introduction to this report, and described in detail in Appendix A, the committee recognizes the limitations of the survey effort underlying the data presented in this chapter. The survey covered 1,400 STTR awards made by the five study agencies during the period FY1998-2010. There was a recorded point of contact for each of the 1,400 awards. This is the preliminary survey population. Awards made in FY2010 would only have started to generate commercial products by the time of the survey in FY2014, so following established practice from prior Academies and GAO surveys, no awards made after FY2010 were surveyed. Given the time period covered, it is unsurprising that many points of contact could not be reached. Of the 1,400 contacts comprising the preliminary survey population, 807 could not be reached, leaving an effective population of 593. Of these, 292 answered questionnaires, generating a response rate of 20.9 percent for the preliminary population and 49.2 percent for the effective population.

The committee acknowledges that because no information was gathered from non-respondents, the data are likely to be biased toward surviving firms. At the same time, the committee notes that successful PIs who left the original firm to start a new venture and successful firms which merged or were bought out by other firms are also excluded from the results. The committee suggests that, where feasible, future assessments of the SBIR program include comparisons of non-awardees, such as in matched samples (Azouley et al., 2014), or regression discontinuity analysis (Howell, 2015). In addition, future assessments should document the root cause of non-responsiveness though this will require agency staffs do a better job of maintaining award databases with routine updates of PI contact information. For example, determining whether the company is still in business even if the PI is no longer with the firm could provide useful evidence about the effectiveness of the SBIR award.

view of company-level activities. In all cases, number of responses (*N*) is provided in each table and figure.

CONNECTING RESEARCH INSTITUTIONS (RIs) TO SMALL BUSINESS CORPORATIONS (SBCs)

STTR has only one congressionally mandated objective: to enhance the connection between SBCs and RIs (see Chapter 1 for a history of the program). Given that STTR projects require an RI as a partner, it is not surprising that the percentage reporting any RI involvement in the STTR project is very high (95 percent) and much higher than for SBIR projects (46 percent) (see Table 5-1).

There were also substantial differences between SBIR and STTR with regard to the kind of university linkage (see Table 5-1):

- Thirty-two percent of STTR respondents reported that the principal investigator (PI) was a university faculty member, compared to only 3 percent of SBIR respondents. This reflects the SBIR mandate that the PI must work more than 50 percent at the SBC, making it much harder to be simultaneously the PI of an SBIR project and a faculty member.
- Seventy percent of STTR projects reported using a faculty member as a subcontractor (compared to 26 percent for SBIR respondents).
- More than twice as many STTR respondents reported that faculty members worked on the project as a consultant (53 percent) as did SBIR respondents (26 percent).
- Similarly, many more STTR projects reported that graduate students worked on the project (51 percent compared to 20 percent for SBIR).
- STTR projects were also much more likely to report that they licensed technology from the RI (18 percent compared to 7 percent for SBIR).
- Finally, STTR projects were also much more likely to report that the project's technology was originally developed at the RI by a project team member (29 percent as compared to 11 percent for SBIR).

TABLE 5-1 STTR Project Connections to Research Institutions (RIs)

	Percentage of Responses	
	STTR Awardees	SBIR Awardees
The PI for this project was at the time of the project an RI faculty member	32.3	2.8
The PI for this project was at the time of the project an RI adjunct faculty member	2.3	5.3
Faculty member(s) or adjunct faculty member(s) worked on this project in a role other than PI	53.0	26.0
Graduate students worked on this project	51.1	20.3
The technology for this project was licensed from an RI	18.4	6.9
The technology for this project was originally developed at an RI by one of the participants in this project	29.3	11.1
An RI was a subcontractor on this project	70.3	25.8
None of the above	4.5	53.7
BASE: TOTAL RESPONDENTS ANSWERING QUESTION	266	1,795

SOURCE: 2011-2014 Survey, Question 71.

STTR also connects SBCs to a considerable number of different RIs. Survey respondents reported 292 projects connected to 167 different RIs.[1] About a quarter of STTR respondents indicated that this was a new connection.[2]

STTR respondents were asked about the impact of the award on their relationship with the RI. Despite the difficulties described by some respondents and interviewees described in Chapter 4, 71 percent of respondents said that the award substantially or somewhat enhanced that relationship, while just 5 percent said that it worsened the relationship (see Table 5-2). Almost three-quarters of STTR respondents indicated that they had a preexisting relationship with the RI already in place, which suggests that creating new relationships is perhaps a less important feature of the program.[3]

Beyond direct connections between RIs and SBCs, there are other metrics that illustrate knowledge effects that derive from the STTR program. Patents and peer-reviewed papers offer an appropriate starting point.

About one-half of STTR respondents claimed that their company was awarded at least one patent related to any STTR- or SBIR-funded technology; 6 percent of these respondents reported at least 10 related patents (see Figure 5-1). Firms reporting on the basis of an SBIR award on average reported more than twice as many patents than those reporting on the basis of an STTR award.

The questionnaire also asked about intellectual property related to the specific award being surveyed. Forty-two percent of STTR respondents reported receiving at least one patent related to the surveyed technology. Two percent reported receiving five or more related patents (see Figure 5-2). There were no significant differences between SBIR and STTR respondents.

In addition to patents, the survey asked about articles published in peer-reviewed journals. Interviews with company executives indicated that, even though technical knowledge and trade secrets are very important, many companies strongly supported peer-reviewed publication. In part, companies saw this as marketing among peers, both for eventual products and a means of

TABLE 5-2 STTR Impacts on SBC-RI Relationships

	Percentage of Responses
Substantially enhanced it	37.8
Somewhat enhanced it	33.2
Made no real difference	23.7
Made it somewhat worse	4.6
Made it substantially worse	0.8
BASE: STTR AWARD RECIPIENTS	262

SOURCE: 2011-2014 Survey, Question 74.

[1]2011-2014 Survey, Question 72. N=167.
[2]2011-2014 Survey, Question 75. N=263.
[3]2011-2014 Survey, Question 75.

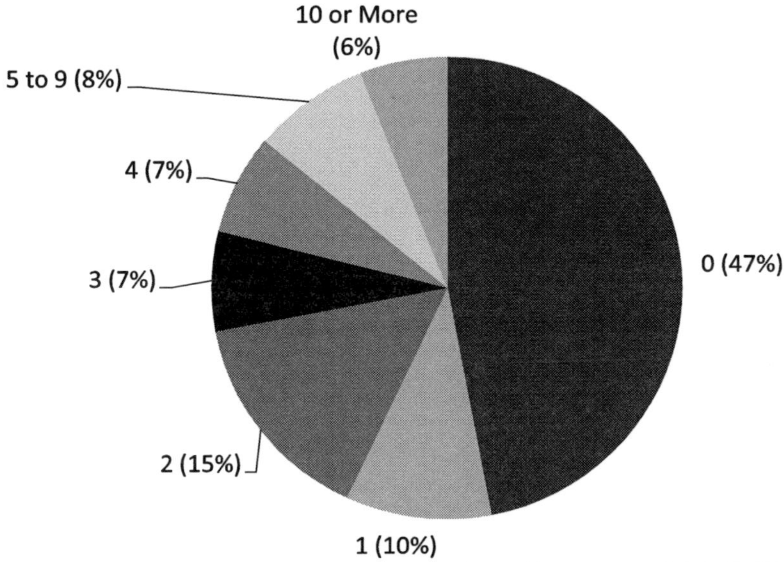

FIGURE 5-1 Number of patents related to all company SBIR/STTR awards (percentage of STTR company-weighted responses).
SOURCE: 2011-2014 Survey, Question 12. N=155.

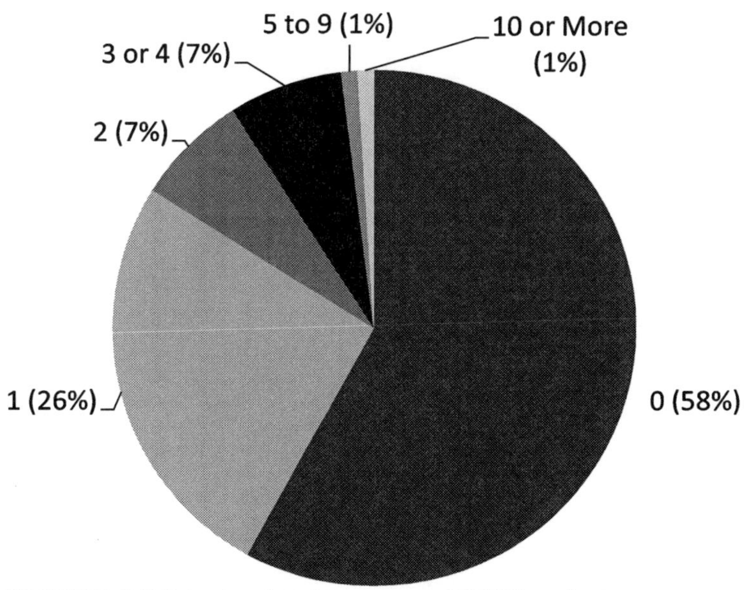

FIGURE 5-2 Patents related to surveyed STTR project.
SOURCE: 2011-2014 Survey, Question 38.1. N=196.

attracting talent. Eighty-two percent of STTR respondents reported publishing at least one peer-reviewed article related to the surveyed project. Forty-six percent reporting publishing at least three such papers (see Figure 5-3).

Conclusions: The STTR Impact

Overall, the university connection is much deeper and richer for STTR awards than for SBIR. This is a central conclusion of the report because it suggests that STTR does provide a different set of outcomes than SBIR and that it does address its specific congressional mandate to stimulate partnerships between SBCs and RIs to an extent that SBIR does not.

Data from the Academies survey shows that STTR companies are especially closely connected to universities. Ninety-five percent of respondents reported a university connection on the surveyed project, a figure far higher than that for SBIR respondents (46 percent); STTR respondents report much higher connection to RIs for every subcategory of connection as well. About one quarter of STTR respondents indicated that the award had helped generate a new connection to an RI, while 71 percent indicated that the award had helped improve an existing connection.

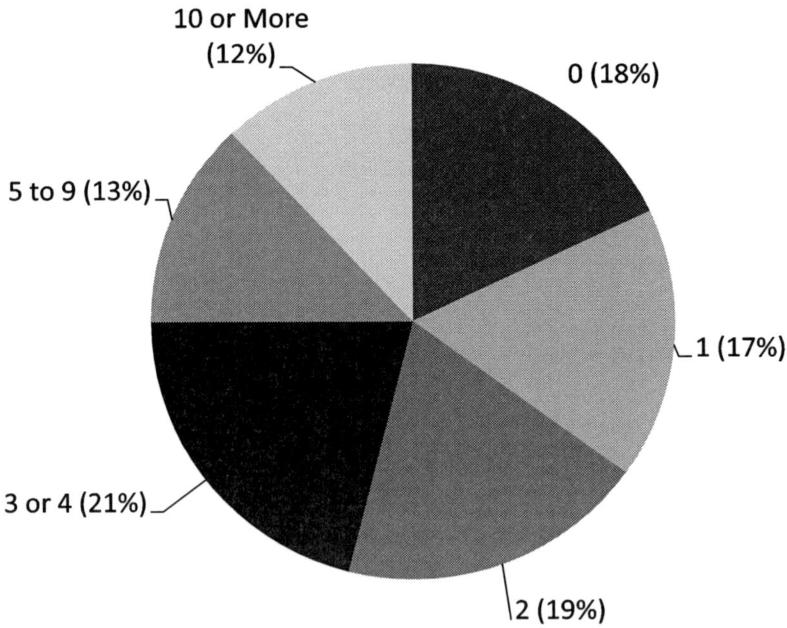

FIGURE 5-3 Number of peer-reviewed articles related to surveyed STTR project (percentage of respondents).
SOURCE: 2011-2014 Survey, Question 39.4.2, N=215.

However, SBCs also report that STTR awards are much harder to manage than SBIR awards: only 3 percent thought STTR awards were easier to manage, while 45 percent thought they were harder.[4]

What emerges from these data is a picture of companies that are dynamic centers of technological innovation, a considerable amount of which is protected through the patent system. About one-half of STTR companies reported that they had received at least one patent based on their work under STTR contracts, while 42 percent reported at least one patent related to the surveyed project only.

STTR companies participate at a high level in the standard forms of technical knowledge dissemination: publishing in peer-reviewed journals and patenting. Eighty-two percent of STTR respondents reported that their company published at least one article based on the STTR-funded work, and 46 percent reported publication of three or more such papers.

COMMERCIALIZATION

As with previous Academies reports on the SBIR program, a broad view of commercialization was adopted, taking it to include additional investments in technology development from outside the SBIR or STTR program as well as sales and licensing revenues. Having determined that no single metric can appropriately capture such a broad concept, and given the long time to market required for many life sciences technologies, a range of benchmarks and metrics have been included.

That said, the focus here is first on different ways of measuring sales and other types of commercial revenue as well as additional investment. In line with previous studies by the Academies and consistent practice at all agencies, additional investment beyond Phase II equates to acknowledgement by third parties that the project has developed technologies of marketable value. For many projects, additional investment is required before commercial sales can begin.

Project-related Revenues

While avoiding overreliance on a single metric, the single most used metric for assessing the STTR and SBIR programs has been project-related revenues.[5]

---

[4]2011-2014 Survey, Question 80. N=211.

[5]For example, National Research Council, *An Assessment of the Small Business Innovation Research Program at the National Institutes of Health*, Washington, DC: The National Academies Press, 2009. Effective July 1, 2015, the institution is called the National Academies of Sciences, Engineering, and Medicine. References in this report to the National Research Council or NRC are used in an historic context identifying programs prior to July 1.

Reaching the Market

The first survey question reported in this section concerns reaching the market: Did the project generate any sales, and if not, are sales expected? (The latter is a necessary add-on question given the long cycle time of some projects.) This question asks about sales and not about other revenues. Responses are summarized in Figure 5-4.

Overall, about 40 percent of STTR projects reported some sales or licensing revenues to date, and a further 28 percent expected sales in the future.

Amount of Sales and Licensing Revenues

The percentage of projects reaching the market is an important metric, but it is not sufficient; it is also necessary to understand the distribution of sales. The 2011-2014 Survey asked those who reported sales of the technology developed for the project to, in addition, report the amount of sales, grouped into ranges. These data are summarized in Figure 5-5.

Most respondents reported sales at the lower end of the scale: 62 percent were less than $500,000. One percent reported sales of at least $20 million, while about 5 percent reported sales of between $5 million and $20 million. [6]

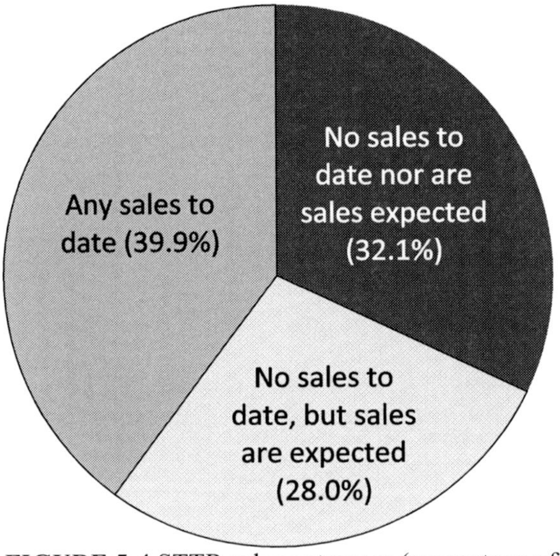

FIGURE 5-4 STTR sales outcomes (percentage of responses).
SOURCE: 2011-2014 Survey, Question 32. N=271. See Table G-7 for details.

[6] Totals in Figure 5-5 do not match Table G-8 exactly because the former were rounded to whole numbers.

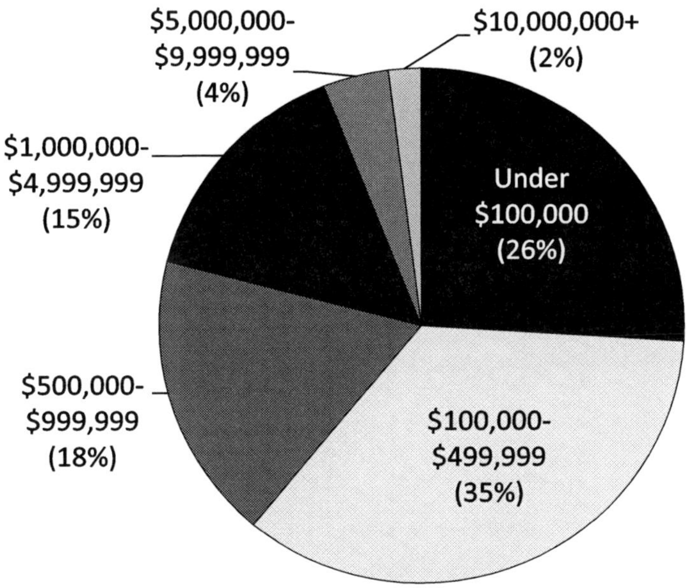

FIGURE 5-5 Distribution of total sales, by dollar range (percentage of responses).
SOURCE: 2011-2014 Survey, Question 34. N=95 (projects reporting sales>$0). See Table G-8 for details.

Markets by Sector

The 2011-2014 Survey asked respondents about the market sectors in which sales were made. Overall, 40 percent identified the domestic private sector, followed by export markets (17 percent). (see Table 5-3).

Additional Investment

The ability of STTR projects and companies to attract additional investments has traditionally been a defining metric for commercialization outcomes.[7] Seventy-one percent of STTR survey respondents indicated that their company received additional investment in the technology related to the surveyed project. As with prior surveys, there is substantial skew with regard to

[7]See National Research Council, *An Assessment of the SBIR Program*, Washington, DC: The National Academies Press, 2008.

TABLE 5-3 Markets for STTR Products and Services

	Percentage of Total Sales	
	STTR Awardees	SBIR Awardees
Domestic private sector	39.8	39.0
Export markets	16.6	11.5
Department of Defense (DoD)	14.4	17.4
NASA	2.0	2.7
Prime contractors for DoD	10.4	10.5
Prime contractors for NASA	0.5	0.9
Agency that awarded the Phase II (if not NASA or DoD)	1.8	1.2
Other federal agencies	3.5	4.2
State or local governments	0.4	2.5
Other	10.6	10.1
BASE: ANY SALES RESULTING FROM THE PROJECT	99	889

NOTE: Respondents were asked to provide a percentage breakdown by market. The table shows the mean of responses for each category.
SOURCE: 2014 Survey, Question 35.

the amount of additional funding received: 46 percent of respondents who received additional investment reported receiving less than $100,000, while slightly less than 1 percent reported more than $20 million in additional funding. (See Table 5-4.)

There has also been interest in the sources of additional funding for high-tech innovation. Although the United States has historically been a leader in venture capital and angel investment, these are not the only or even the primary sources of additional funding for STTR projects (see Table 5-5). Of those STTR projects that reported additional funding (N=189), 54 percent reported additional funding from internal sources, 46 percent from non-SBIR/STTR federal sources, and 32 percent from other external sources such as state governments and foundations. Fifteen percent reported investments from strategic partners, six percent reported venture capital funding, and 3 percent reported funding from angel and other private equity investors. (See Table 5-5.)

COUNTERFACTUALS

Because there is no available matched set of companies that did not receive STTR funding at precisely the point in time that surveyed companies did receive Phase II funding, it is not possible to develop an appropriate control group against which to measure impacts (see the discussion of Academies efforts to do so in Appendix A).

TABLE 5-4 Additional Funding by Program and Amount

	Percentage of Responses	
	STTR Awardees	SBIR Awardees
None ($0)	29.2	31.3
Under $100,000	46.2	47.1
$100,000-$499,999	9.5	8.5
$500,000-$999,999	6.1	4.3
$1,000,000-$4,999,999	7.2	6.2
$5,000,000-$9,999,999	0.8	1.1
$10,000,000-$19,999,999	0.4	0.7
$20,000,000-$49,999,999	0.8	0.6
$50,000,000 or more		0.2
Mean (Thousands of Dollars)	691.7	832.1
Median (Thousands of Dollars)	50	50
BASE: TOTAL RESPONDENTS ANSWERING QUESTION	264	1,827

NOTE: STTR and SBIR data are aggregated over all five agencies for FY 1998–2007 for DoD, NSF, and NASA and FY2001–2010 for NIH and DoE, inclusive.
SOURCE: 2011-2014 Survey, Question 30.

However, it is at least possible to ask—as previous Academies and Government Accountability Office (GAO) surveys have done—what the company itself believed might have happened had STTR funding not been available. Although this is of course subjective, the company is best suited to provide these answers.

Because alternative funding especially for long-cycle projects is difficult to acquire, it is not surprising that only 9 percent of respondents believed that the project would definitely or probably have gone ahead without funding. Conversely, three-quarters of respondents said that the project would probably or definitely not have proceeded.[8]

These data have interesting wider implications for debates about early-stage funding; notably, they suggest weak support for the "crowding out" hypothesis (that public funding displaces private investment). Awardees in our survey—presumably those with the closest knowledge of funding prospects for the project—overwhelmingly believed it to be unlikely that alternative private funding would be found. These results also underscore the importance of SBIR/STTR funding for these small companies.

[8]2011-2014 Survey, Question 24.

TABLE 5-5 Distribution of Additional Investment by Source of Funds

	Percentage of Responses	
	STTR Awardees	SBIR Awardees
Non-SBIR/STTR federal funds	45.5	39.5
Private Investment: U.S. Sources	30.2	35.6
Venture capital (VC)	6.3	5.4
U.S. angel funding or other private equity investment (not VC)	3.3	8.4
Friends and family	5.3	4.3
Strategic investors/partners	15.3	13.1
Other sources	11.1	15.3
Foreign Investment	2.6	4.0
Other External Sources	*32.3*	*14.2*
State or local governments	16.9	10.7
Research institutions (such as colleges, universities or medical centers)	17.5	4.4
Foundations	3.2	1.0
Internal Sources	*54.0*	*69.9*
Your own company (Including money you have borrowed)	45.5	62.1
Personal funds	14.3	17.3
BASE: TOTAL RESPONDENTS ANSWERING QUESTION	189	1,239

NOTE: Responses for subcategories do not total to categories because more than one response was permitted.
SOURCE: 2011-2014 Survey, Question 31.

COMPANY IMPACTS

Although the effect of STTR funding on the company is not directly included in the congressional objectives for the program, helping small companies to become self-sufficient (and in some cases to grow rapidly) has implications for program impacts and is therefore included in this analysis. Thirty-six percent of STTR respondents reported that the award had at least in part supported company formation.

Small high-tech companies are often fluid in structure, and the 2011-2014 Survey found that many participating companies had changed structurally in recent years. Sixteen percent had established strategic partnerships with major players, while 12 percent had spun-off at least one company and the same percentage had been acquired by or merged with another firm.

Ideally, companies that receive SBIR/STTR funding become more stable and develop other contracts and revenues over time. This appears to be the case for STTR recipient companies because dependence on SBIR/STTR funding is limited. Overall, 48 percent of STTR respondents indicated that SBIR/STTR funded 10 percent or less of total R&D effort, while 28 percent overall indicated that it funded greater than one-half. This picture is reinforced by data on sources of company revenues. Thirty-one percent of STTR companies reported zero SBIR/STTR revenues, while about 24 percent reported receiving greater than one-half of the company's revenues from SBIR/STTR in their most recent fiscal year.

The survey also asked about the overall impact of SBIR/STTR on the company. As Figure 5-6 shows, 32 percent of survey respondents saw a highly positive or transformative effect, and another 46 percent reported a positive impact.

PARTICIPATION OF WOMEN AND MINORITIES

The SBIR program is mandated by Congress to foster the participation of women and minorities. As noted in Chapter 1, the STTR program is not subject to such a mandate. However, we believe that it is still useful to determine levels of participation for women and minorities in the STTR program at the five study agencies. In following the methodologies developed for previous assessments of the agency SBIR programs, the Committee

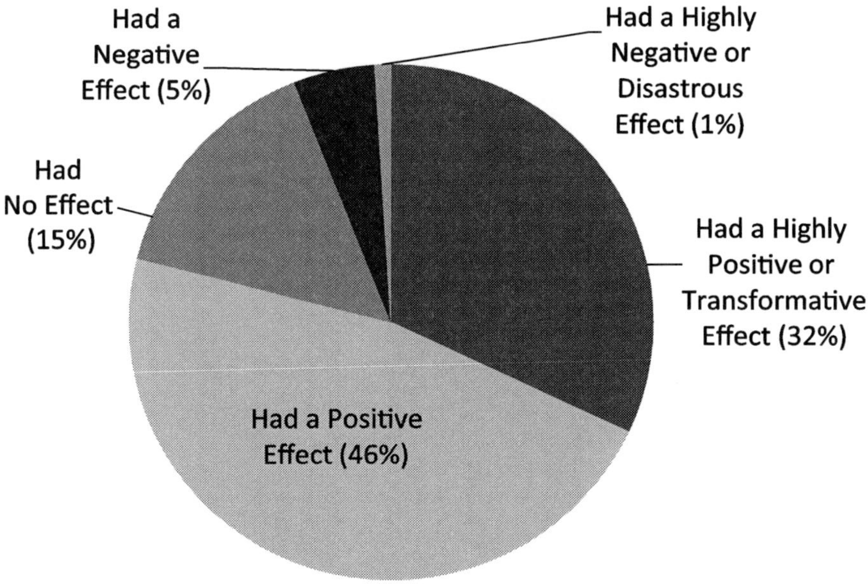

FIGURE 5-6 Long-term impact of SBIR/STTR on companies.
SOURCE: 2011-2014 Survey, Question 57. N=267.

reviewed survey data to examine the participation of women and minorities through company ownership and as principal investigators (PIs).

Awards data from the Small Business Administration (SBA) are not included in this assessment because a comparison between selected agency data and SBA data indicated that the latter does not reliably report company ownership.

Participation of Woman-owned Firms

Figure 5-7 shows that woman-owned firms accounted for 8 percent of the respondent firms for all five study agencies.

This percentage was lower than the 10.6 percent for SBIR.

Woman Principal Investigators

Because PIs are an important part of the pipeline that eventually leads to scientists and engineers forming their own companies, the Academies survey sought information about the percentage of PIs who were women (see Figure 5-8).

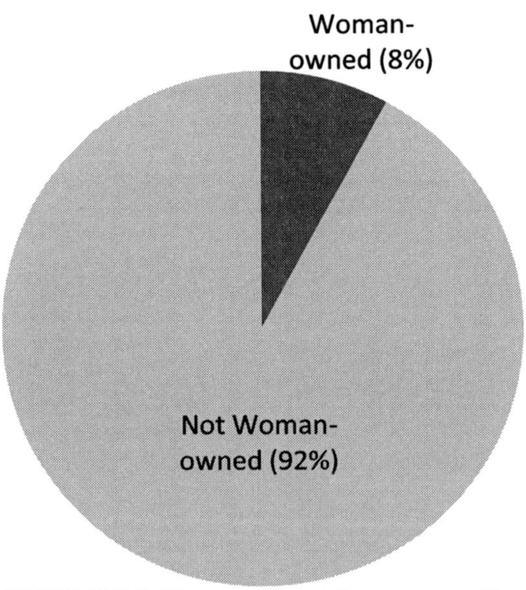

FIGURE 5-7 Woman-owned firms among Phase II STTR projects at five study agencies, 1997-2010 (percentage of responses).
SOURCE: 2011-2014 Survey, Question 15. N=263.

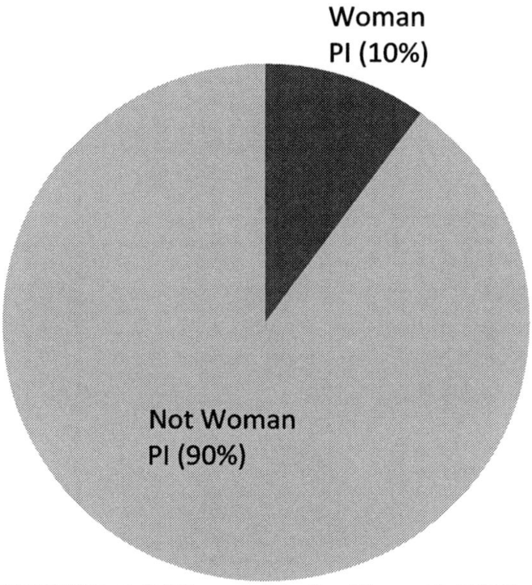

FIGURE 5-8 Woman PIs for Phase II STTR projects at five study agencies, 1997-2010 (percentage of responses).
SOURCE: 2011-2014 Survey, Question 16. N=270.

Survey respondents reported that women constituted 10 percent of STTR PIs, which is slightly higher than the percentage reported for SBIR (8.9 percent).

Minority-owned Firms

The Academies survey also addressed the question of minority-owned firms. Overall, survey respondents reported that 9.1 percent of STTR firms were minority-owned at the time of the award. This was the same percentage reported for SBIR. Following the methodology established in previous reports, the survey also asked about the minority affiliation of the ownership (see Table 5-6).

The percentages for STTR are closely similar to those for SBIR. In both cases, the number of firms owned by Black, Hispanic, and Native American owners is extremely small: among STTR companies the survey identified three Black-owned firms, three Hispanic-owned firms, and one Native American-owned firm.

Minority Principal Investigators

The Academies survey also inquired about the incidence of minority PIs. Table 5-7 shows that about 14 percent of respondents indicated that surveyed project had a minority PI but that a large majority of these were Asian-Americans.

A NOTE ON TALENT PATHWAYS

Woman- and minority-owned firms result from a series of pathways that stretch back to the varied participation of students in K-12 in science-oriented programs, through college-level choices, graduate schools, post-doctoral programs, and then employment in science and engineering fields.

TABLE 5-6 Minority-Owned Firms: Participation in STTR, 1998-2011

	Percentage of Responses	
Minority Ownership	STTR Awardees	SBIR Awardees
Asian Indian	4.6	3.5
Asian Pacific	2.3	3.4
Black	1.1	0.3
Hispanic	1.1	1.5
Native American	0.4	0.2
Other	0.4	0.3
Not minority	89.9	89.9
N=	263	1,789

NOTE: Percentages do not sum to 9.1 because multiple responses were permitted.
SOURCE: 2011-2014 Survey, Question 15.

TABLE 5-7 Minority PIs for STTR/SBIR Awards, 1997-2011

	Percentage of Responses	
	STTR Awardees	SBIR Awardees
Asian Indian	4.1	4.1
Asian Pacific	4.1	5.1
Black	0.7	0.2
Hispanic	3.0	1.3
Native American	0.7	0.2
Other	1.5	0.5
Not a minority	85.9	88.7
N=	270	1,856

SOURCE: 2011-2014 Survey, Question 16.

At each stage, the population of women and minorities changes, as it does when considering the individual pathways for specific disciplines. For example, women account for more than 50 percent of PhDs in life sciences, but their numbers have recently been falling from already low levels in computer science PhD programs.

From the STTR programs' perspective, a key question is whether participation reflects the distribution of the available talent: whether the numbers reflect the current distribution of potential applicants from woman- and minority-owned firms, and whether the current selection procedures are fair. Reviewing applications and awards data is an important exercise in determining the latter. Other reports have recommended that agencies (particularly NSF) work with experts to find new methodologies for measuring the universe of potential applicants.[9]

---

[9]See National Science Board, *Revisiting the STEM Workforce, A Companion to Science and Engineering Indicators 2014*, Arlington, VA: National Science Foundation (NSB-2015-10), 2015. See also National Research Council, *Expanding Underrepresented Minority Participation: America's Science and Technology Talent at the Crossroads*, Washington, DC: The National Academies Press, 2011.

6

Findings and Recommendations

The findings and recommendations in this chapter reflect the performance of the STTR program against its congressional objective. According to the SBA's STTR Policy Directive:

"The statutory purpose of the STTR Program is to stimulate a partnership of ideas and technologies between innovative small business concerns (SBCs) and Research Institutions through Federally-funded research or research and development (R/R&D). By providing awards to SBCs for cooperative R/R&D efforts with Research Institutions, the STTR Program assists the small business and research communities by commercializing innovative technologies."[1]

FINDINGS

General conclusions about the STTR program must be viewed with caution. STTR programs are managed and operated differently by each agency and in some cases differently by separate components within the Department of Defense (DoD) and the National Institutes of Health (NIH). Therefore, individual agencies will need to view the findings and recommendations provided herein within the specific context of their own programs. Not all findings will be relevant to all agencies.

In our view, four core questions surround the STTR program:

1. Does STTR meet its congressionally mandated objective to improve linkages between small business concerns (SBCs) and research institutions (RIs), and in the process support the enhanced transfer of technology from the latter?

[1]Small Business Administration, Office of Investment and Innovation, "Small Business Technology Transfer (STTR) Program—Policy Guidance," updated February 24, 2014.

2. Does the program meet other objectives, notably:

   - Support of agency mission, including the direct provision of technologies for use by the acquisition agencies;
   - Enhanced participation by women and minorities; and
   - Development of innovative small companies?

3. Does the program provide benefits that are not available—or to a significant degree not as available—through the SBIR program and other funding pathways?
4. What best practices can be identified, and in what areas can the operations of the program be improved?

Sources of Findings

The committee's findings are based on a range of tools, including a survey of award recipients, case studies of awardee companies, agency data, public workshops, agency meetings, and reports and presentations submitted by the agencies. Information on which to assess these aspects of the program has been drawn from the Academies 2011-2014 Survey, which is described in detail in Appendix A and in Appendix C, company case studies profiled in Appendix E, discussions with university technology transfer officials, a series of ongoing discussions and conversations with agency officials, and the workshop convened by the committee on the STTR program in Washington, DC on May 1, 2015.

The Academies 2011-2014 Survey was sent to every principal investigator (PI) in the population who received a Phase II award from NIH and DoE in FY2001-2010 and from NSF, NASA, and DoD in FY1998-2007. The preliminary population prior to contact was 1,400. Of these, 807 principal investigators were determined to be not contactable at the STTR company listed in the agency awards databases. The remaining 593 awards with their prospective principal investigator contacts constitute the population for this study. From these, 292 responses were received, for a preliminary population response rate of 20.9 percent and a population response rate of 49.2 percent.

A. STTR is meeting its congressional objective of fostering cooperation between small business concerns and research institutions, and does so in some respects to an extent that SBIR does not.

1. Overall, the university connection is much deeper and richer for STTR awards than for SBIR, and STTR addresses its congressional mandate

to stimulate partnerships between small business concerns and research institutions to an extent that SBIR does not.[2]

2. STTR projects generate wider and deeper linkages between small businesses and research institutions than do SBIR projects, according to data from the Academies 2011-2014 Survey.[3] Most notably:

    - Thirty-two percent of STTR respondents reported that the principal investigator was a university faculty member, compared to only 3 percent of SBIR respondents, across the five major agencies.
    - More than half of STTR respondents reported graduate students working on the project, compared to 20 percent for SBIR.
    - STTR respondents were far more likely than SBIR respondents (18.4 percent vs. 6.9 percent) to report that the technology underpinning the project was licensed from or originated at the research institution.
    - Seventy percent of projects reported that a faculty member at a research institution was a consultant on the project, compared to 26 percent for SBIR.
    - Case studies of STTR companies offer several examples of successful collaborations.[4] They illustrate a number of different ways in which STTR works to

        - Create deeper links between the research institution and the small business concern;
        - Allow the small business concerns to enter technically new or challenging areas;
        - Identify and collaborate with possible recruits, especially graduate students; and
        - Access research institution equipment and expertise, as well as access alumni and other social networks, including angel investors and alumni commercialization funds.

3. STTR projects fully participate in the dissemination of knowledge through patents and publications.

    - Forty-two percent of STTR respondents reported at least one patent, almost as many (44 percent) as those reported by SBIR respondents.[5]

---

[2]See Chapter 5 (Quantitative Outcomes).
[3]See Table G-1.
[4]See Appendix E (Case Studies). See also the discussion in Chapter 4 (Qualitative Assessment: Company and University Perspectives).
[5]See Table G-19.

- STTR respondents reported a higher percentage of participation than SBIR respondents in publishing peer-reviewed papers, being more likely to produce at least one peer-reviewed paper and also more likely to produce a larger number of papers.[6]

B. Perspectives on STTR use and management vary by agency. Some see it as a link between basic research and acquisition programs; others see STTR as having similar objectives to SBIR and therefore operate the two programs in tandem.[7]

1. Program managers at NASA and at DoD (in particular the Army and Navy) see STTR as filling a gap between basic research and acquisition programs.

- STTR is now used by the Navy and Army as a means of addressing potentially valuable technologies at lower technology readiness levels (TRLs) that are not necessarily aligned immediately with the needs of the acquisition programs. In effect, STTR is used to undertake preliminary work on technologies that may eventually become ready for acquisition. It overlaps with the TRLs served by SBIR, but expands the range to include the earliest technology readiness level.
- This view may now carry more weight as the SBIR program is increasingly aligned with the immediate needs of the acquisition programs at DoD.[8] As a result, SBIR awards have become increasingly focused on serving the immediate needs of the Army, Navy and other components, and in particular their acquisition offices.[9]
- NASA also uses different mechanisms to develop topics for STTR, and, like DoD, places the program strategically as focused on lower TRL levels than SBIR.
- As a result, NASA and DoD program managers view the STTR program as a significant conduit between their agency and leading research universities. Program managers report that the STTR program provides a mechanism through which these procurement agencies can explore cutting-edge technologies in a research environment.

---

[6]See Table G-20.
[7]See Chapter 2 (Program Management).
[8]See National Research Council, *SBIR at the Department of Defense*, Washington, DC: The National Academies Press, 2014, Chapter 2.
[9]Efforts to interview Air Force STTR personnel were not successful.

2. Program managers at NIH, NSF, and DoE do not see an additional value in the STTR program. They see STTR as having similar objectives to SBIR and therefore operate the two programs in tandem.

   - All three agencies use the same solicitation for SBIR and STTR.
   - All use the same selection processes and procedures.
   - Pilot efforts at NSF to differentiate SBIR and STTR through different topics failed, and the agency has ended the pilot.
   - DoE allows small businesses to apply simultaneously for SBIR and STTR, indicating that the agency sees essentially no difference between the programs.

C. To a considerable extent, STTR fosters private-sector commercialization of innovations derived from federal R&D.

- Forty percent of STTR projects reported that they had made some sales, and a further 28 percent expected to do so in the future.[10] In comparison, 49 percent of SBIR projects reported sales.
- For most of the ranges of responses, the size of reported commercialization was broadly similar for SBIR and STTR, although the former reported a higher percentage of large winners.[11]
- While technologies created by SBIR and STTR programs serve similar markets, STTR projects were more likely to serve export markets.[12]
- Seventy-one percent of STTR respondents indicated that their projects received additional investment after the Phase II STTR award. Nine percent reported receiving $1 million or more (the same percentage as SBIR).[13]

D. The participation of women and minorities in the STTR program is low and not actively fostered.[14]

1. Data from the Academies 2011-2014 Survey indicate that the participation of women in the STTR program is low.[15]

   - Survey respondents reported that woman-owned firms accounted for 8 percent of all STTR Phase II firms.

---

[10]See Table G-7. The committee adopts a broad view of commercialization, taking it to include additional investments in technology development from outside the SBIR or STTR program as well as sales and licensing revenues. See Chapter 5 (Quantitative Outcomes) and Appendix G (Annex to Chapter 5: Quantitative Outcomes).
[11]See Table G-8.
[12]See Table 5-3 and Table G-9.
[13]See Table G-12.
[14]See Chapter 5 (Quantitative Outcomes).
[15]See Figures 5-7 and 5-8.

- Women accounted for 10 percent of STTR principal investigators. Although the percentages of women in some STEM fields have been declining recently, this is not the case in others (e.g., life sciences).

2. Data from the Academies 2011-2014 Survey indicate that the participation of Black, Hispanic, and Native Americans in the STTR program is extremely low.[16]

   - Overall figures for minority participation using the SBA definition remain low. The Academies 2011-2014 Survey reported that 9.1 percent of STTR respondents were from minority-owned firms.
   - The survey also found that 14.1 percent of principal investigators were reported to be from socially or economically disadvantaged groups. These included 0.7 percent Black, 3 percent Hispanic, and 0.7 percent Native American.
   - The SBA definition of socially or economically disadvantaged groups is inadequate to reflect congressional objectives. Data reported by the agencies obscures the extremely low level of participation from other disadvantaged groups by including Asian Americans. The Academies survey found that only 1.1 percent of respondents were from Black- and Hispanic-owned firms respectively, and 0.4 percent was from Native American-owned firms.

3. Agencies have no outreach efforts or other programs designed to foster such participation specifically in their STTR program.

E. **STTR is aligned with agency missions and the take-up of technologies within acquisition agencies.**

1. As described in Chapter 2, topics for STTR solicitations are developed through processes that are either identical to or parallel those used for SBIR awards.
2. STTR project selections are made with a close view to agency mission.

   - For the acquisition agencies (DoD and NASA) STTR projects are selected in large part because their results meet specific technology needs of these agencies.[17]
   - For NSF, DoE, and NIH—agencies that do not usually procure the technologies developed by STTR—STTR project selections are

[16]See Tables 5-6 and 5-7.
[17]See Chapters 2 (Program Management) and 4 (Qualitative Outcomes: Company and University Perspectives).

made based on the same criteria used in making all agency awards (including non-SBIR/STTR), which closely reflect the agencies' missions.[18]

- Many STTR projects, particularly those funded by DoE, support partnerships with the National Laboratories.[19]
- Case studies (as well as success stories published by the agencies) support the evidence that STTR contributes to meeting specific agency needs.[20]

F. STTR awards require a formal partnership between the small business concern and the research institution, but they each can have different interests and needs. This creates unique challenges within the STTR program.

1. In particular, research institutions see the development and widespread dissemination of technical knowledge as part of their core mission. In contrast, small businesses see the commercialization of knowledge as a priority, which will likely require steps to limit the ability of others to use technical information, through the use of either trade secrets or patents.
2. As university faculty participate in commercial activities outside the research institution, university administrators often seek to ensure that a dividing line exists between research inside the university and activities outside.

   - They often require that the faculty be a full-time employee of the university, which means that the faculty member cannot work more than half-time for the small business concern. STTR—but not SBIR—can accommodate principal investigators who are still working more than 50 percent time at the RI.
   - Conflict of interest (COI) rules may require that small businesses have principal owners/managers other than the founding faculty member, thus significantly reducing incentives for faculty participation.
   - COI rules may also block the employment of graduate students at the small business concern.
   - COI rules may also limit or prevent use of university laboratories and facilities acting as subcontractors to a small business concern that is wholly or partly owned by a faculty member.

---

[18]See Chapter 2 (Program Management).
[19]See Chapter 4 (Qualitative Outcomes: Company and University Perspectives).
[20]See Appendix E (Case Studies) and Chapter 4 (Qualitative Outcomes: Company and University Perspectives).

3. Research institutions have varied views on and approaches to licensing of university intellectual property (IP).

    - Typically, technologies at issue in STTR projects are at a very early stage of development when it is not clear how commercially successful they will be or what the size of the relevant market will be. Yet some universities require significant royalty payments upfront, as well as payment for patenting costs and other expenses. This approach to licensing makes it difficult for small businesses to work with these universities and makes a successful partnership through STTR less likely.
    - Some universities have developed an approach to licensing intellectual property that is more supportive of STTR/SBIR activity. The University of Minnesota, for example, has developed two well-defined pathways to licensing that appeal to small companies as evidenced by the substantial increase in licensing and partnership agreements since the policy was introduced. The technology transfer offices (TTOs) at the University of New Hampshire and North Carolina State University, for example, each see their mission as calling for them to support STTR/SBIR activity while still managing COI issues. Progress is also being made in this regard by university groups such as AUTM, NACUBO, and NACUA.

4. The bureaucracy at research institutions can be challenging. Research institutions are big organizations, typically with large overhead rates, and the transfer of technology often is not seen as a core part of their mission. Unless there is a defined path to partnership, negotiating with research institutions can take considerable time and resources for a small business.

    - Agreements with research institutions may require multiple levels of approval.
    - Adjustments and changes—often necessary for very early stage technology projects—may require further approvals and permissions, which can cause substantial delays.
    - The overhead costs at research institutions may be viewed as an unproductive burden from the perspective of the small business concern.

G. Small business concerns in general see STTR as more onerous to use and thus less attractive than SBIR, in part because STTR awards

require a formal partnership between the small business and the research institution.[21]

1. About 45 percent of small business survey respondents indicated that the STTR program was more difficult to use than SBIR. Only 3 percent thought the opposite.[22]

    - This view is reflected in some case studies where the principal investigators indicated that they applied for STTR only when absolutely necessary. One experienced principal investigator said that given the choice he would utilize SBIR over STTR "every time."
    - Many case study companies and survey respondents indicated that the success of the STTR partnership depended heavily on the degree of commitment of the RI, which varied widely.

2. Small business concerns see STTR partnerships with the National Laboratories as especially challenging.[23]

    - National Laboratory administrators consider STTR to be a small amount of money and a considerable amount of work.
    - There is little evidence that National Laboratories see STTR as a strategic solution to disseminating their technology.
    - Small business concerns expressed considerable frustration with the difficulties of holding National Laboratories (as well as other research institutions) accountable for STTR deliverables, something that was less the case for SBIR.
    - Small business concerns also expressed concern about the cultural differences between the open culture of some laboratories and the more closed commercial culture of private businesses. This problem could be resolved, but in some cases can lead to difficulties for the small business.
    - However, case studies indicate that STTR may be a useful mechanism through which National Laboratory scientists and engineers can pursue applications of advanced ideas.

H. STTR supports the development of innovative companies.

1. Survey respondents report that STTR is having a transformative or strongly positive impact on participating companies, with 32 percent of

---

[21]See Chapter 4 (Qualitative Outcomes: Company and University Perspectives) and Appendix E (Case Studies).
[22]See Figure G-1.
[23]See Chapter 4 (Qualitative Outcomes: Company and University Perspectives).

those surveyed reporting that the award has a "highly positive transformative" effect on the company, and an additional 46 percent reporting that the effect was positive.[24]

2. Case studies indicate that in some cases, STTR provided a bridge between academia and commercialization that would not otherwise exist (see for example the Stratatech case study).

RECOMMENDATIONS

The committee finds that STTR meets the specific congressional objective of increasing the linkages between small business concerns and research institutions. As shown in Table G-3 (page 274), a total of 71 percent of the small business survey respondents indicated that the STTR program enhanced their relationships with the research institutions. However, as shown in Figure G-1 (page 275) 45 percent of the survey respondents found the STTR program to be harder to manage than the SBIR program. This finding is not surprising since STTR requires a formal agreement between the small business concern and the research institution. This requirement, which was originally intended to encourage more small businesses to collaborate with research institutions, may sometimes impede this collaboration. To address this issue, the committee recommends:

A. The five sponsoring agencies should address the following factors that may be discouraging some small businesses and research institutions from collaborating in the STTR and SBIR programs:

- **<u>STTR Program</u>**

  1. **Finding alternative templates for royalties and licensing agreements.** The complexity and variation in intellectual property terms and conditions among universities and laboratories can cause delays in developing contractual agreements between the research institution and small business concerns. The potential partners in an STTR award should consult leading research institutions to learn what templates for royalty and licensing schemes have proven to be most effective and might be adapted for their project. Many different schemes have been used and should be reviewed in the context of the potential project and its participants. One example, adopted at the University of Minnesota, offers a standard option along with an alternate "open negotiation" option that could be a useful template for some projects.

[24]See Table G-26.

2. **Resolving unique challenges of cooperation.** Given the highly flexible nature of the STTR program, the sponsoring agencies should consider seeking SBA authority to act in special circumstances to protect participants from the effects of unexpected delays or related problems with contract agreements or deliverables.
3. **Maintaining a distinct strategy for STTR.** Each sponsoring agency should seek ways to ensure that the STTR program plays an identifiable role in the agency's R&D strategy that differs from that played by the SBIR program. A focus on projects with earlier technology readiness levels might be part of this strategic distinction.

- <u>SBIR Program</u>

4. **Relaxing the small business employment requirement.** Research institutions with personnel who seek to serve as Principal Investigators on SBIR awards while retaining their full-time positions might be allowed—under exceptional circumstances—to seek a waiver of the SBIR 51 percent small business employment requirement.
5. **Reporting on waiver requests.** If any waivers are to be considered, the sponsoring agencies should develop an appropriate mechanism for addressing these special requests and should report on the number of waiver requests and the number granted, as part of their annual program reporting.

- The overall impact of these proposed changes should be evaluated in future assessments of the SBIR and STTR programs to determine if they have been effective in strengthening the collaboration between small business concerns and research institutions in the STTR and SBIR programs.

B. SBA should change its definitions to address congressional intent with regard to minorities.[25]

1. SBA translates "minorities" in the governing legislation into "socially and economically disadvantaged groups" in the Policy Guidance for SBIR. Asian Americans are designated as one of the included groups.
2. Asian Americans are well represented as founders of innovative small businesses. Research shows that they have in recent years accounted for

[25]See Finding D.

a significant number of all startups in Silicon Valley and other innovation clusters.[26]

3. Including Asian Americans has the direct effect of underplaying the low participation for African American, Hispanic American, and Native American entrepreneurs and principal investigators.
4. SBA should act immediately to change its definitions to ensure that efforts in this area are focused on activities that meet congressional intent.
5. SBA should also require that agencies collect data—and report annually—on the participation of each SBA subgroup in the SBIR and STTR programs.

C. Other recommendations[27]

1. **Enhance the participation of women and minorities:** STTR administrative funding should be aligned with SBIR funds for the purpose of enhancing the participation of women and minorities. Given the small size of STTR programs, these administrative funds should be used in joint programs with SBIR to address this issue through enhanced targeted outreach programs.
2. **Streamline connections with the National Laboratories:** DoE should establish a pilot program that streamlines the use of STTR in connection with the National Laboratories. The agency might, for example, consider ways to reduce the multiple layers of permission required for project changes, and to provide improved incentives for National Laboratories to participate. The National Laboratories should have the discretion to be more innovative and take greater risks with the STTR program, given its history of success.
3. **Improved data collection.** More effective management of the STTR program depends on better data collection about outcomes. While current SBA efforts in this area show some promise, we would urge agencies to ensure that they do not delay improved data collection in the interim, and that they ensure that the specific metrics most appropriate for the agency are addressed—for example, at NIH data related to clinical trials and the FDA will be needed.
4. **Additional analysis of National Laboratories role in technology transfer to small business.** It is apparent that working with the National Laboratories is challenging for many small businesses. It would therefore be useful to study this issue further with a view to

[26]See, for example, Anuradha Basu and Meghna Virick (2015), "Silicon Valley's Indian diaspora: networking and entrepreneurial success," *South Asian Journal of Global Business Research*, 4(2):190-208.
[27]See Findings D, E, and G.

developing recommendations related to best practices for this type of research.

APPENDIXES

Appendix A

Overview of Methodological Approaches, Data Sources, and Survey Tools

This series of reports on the Small Business Innovation Research (SBIR) and the Small Business Technology Transfer (STTR) programs at the Department of Defense (DoD), National Institutes of Health (NIH), National Aeronautics and Space Administration (NASA), Department of Energy (DoE), and National Science Foundation (NSF) represents a second-round assessment of the program undertaken by the National Academies of Sciences, Engineering, and Medicine.[1] The first-round assessment, focusing on SBIR and conducted under a separate ad hoc committee, resulted in a series of reports released from 2004 to 2009, including a framework methodology for that study and on which the current methodology builds.[2]

The current study is the first to focus on the STTR program, and it addresses the twin objectives of assessing outcomes from the STTR program and of providing recommendations for improvement.[3] Section 1c of the Small

[1]Effective July 1, 2015, the institution is called the National Academies of Sciences, Engineering, and Medicine. References in this report to the National Research Council or NRC are used in an historic context identifying programs prior to July 1.

[2]National Research Council, *An Assessment of the Small Business Innovation Research Program: Project Methodology,* Washington, DC: The National Academies Press, 2004.

[3]The methodology developed as part of the first-round assessment of the SBIR program also identifies two areas that are excluded from the purview of the study: "The objective of the study is *not* to consider if SBIR should exist or not—Congress has already decided affirmatively on this question. Rather, we are charged with providing assessment-based findings of the benefits and costs of SBIR . . . to improve public understanding of the program, as well as recommendations to improve the program's effectiveness. It is also important to note that, in accordance with the Memorandum of Understanding and the Congressional mandate, the study will *not* seek to compare the value of one area with other areas; this task is the prerogative of the Congress and the Administration acting through the agencies. Instead, the study is concerned with the effective review of each area." National Research Council, *An Assessment of the Small Business Innovation Research Program: Project Methodology*. In implementing this approach in the context of the current round of SBIR assessments, we have opted to focus more deeply on operational questions.

Business Administration (SBA) STTR Directive states program objectives as follows: "The statutory purpose of the STTR Program is to stimulate a partnership of ideas and technologies between innovative small business concerns (SBCs) and Research Institutions through Federally-funded research or research and development (R/R&D)."[4]

SBA also provides further guidance on its web site, which aligns the objectives of STTR more closely with those of SBIR: "(1) stimulate technological innovation, (2) foster technology transfer through cooperative R&D between small businesses and research institutions, and (3) increase private-sector commercialization of innovations derived from federal R&D."[5]

The STTR program, on the basis of highly competitive solicitations, provides modest initial funding for selected Phase I projects (in most cases up to $150,000) and for feasibility testing and further Phase II funding (in most cases up to $1.5 million) for qualifying Phase I projects.

DATA CHALLENGES

From a methodology perspective, assessing this program presents formidable challenges. Among the more difficult are the following:

- **Lack of data.** The agencies have only limited ability to track outcomes data, both in scope (share of awards tracked) and depth (time tracked after the end of the award). There are no published or publicly available outcomes data.
- **Intervening variables.** Analysis of small innovative businesses suggests that they are often very path dependent and, hence, can be deflected from a given development path by a wide range of positive and negative variables. A single breakthrough contract—or technical delay—can make or break a company.
- **Lags.** Not only do outcomes lag awards by a number of years, but also the lag itself is highly variable. Some companies commercialize within 6 months of award conclusion; others take decades. And often, revenues from commercialization peak many years after products have reached markets.

ESTABLISHING A METHODOLOGY

The methodology utilized in this second-round study of the SBIR-STTR programs builds on the methodology established by the committee that completed the first-round study.

[4]Ibid., p. 3.

Publication of the 2004 Methodology

The committee that undertook the first-round study and the agencies under study acknowledged the difficulties involved in assessing the SBIR-STTR programs. Accordingly, that study began with development of the formal volume on methodology, which was published in 2004 after undergoing the standard Academies peer-review process.[6]

The established methodology stressed the importance of adopting a varied range of tools based on prior work in this area, which meshes with the methodology originally defined by the first study committee. The first committee concluded that appropriate methodological approaches

> build from the precedents established in several key studies already undertaken to evaluate various aspects of the SBIR/STTR. These studies have been successful because they identified the need for utilizing not just a single methodological approach, but rather a broad spectrum of approaches, in order to evaluate the SBIR/STTR from a number of different perspectives and criteria.
>
> This diversity and flexibility in methodological approach are particularly appropriate given the heterogeneity of goals and procedures across the five agencies involved in the evaluation. Consequently, this document suggests a broad framework for methodological approaches that can serve to guide the research team when evaluating each particular agency in terms of the four criteria stated above.[7]

Table A-1 illustrates some key assessment parameters and related measures to be considered in this study.

The tools identified Table A-1 include many of those used by the committee that conducted the first-round study of the SBIR-STTR programs. Other tools have emerged since the initial methodology review.

Tools Utilized in the Current STTR Study

Quantitative and qualitative tools being utilized in the current study of the STTR program include the following Academies activities:

- **Surveys.** An extensive survey of STTR award recipients as part of the analysis.

[6]National Research Council, *An Assessment of the Small Business Innovation Research Program: Project Methodology*, p. 2.
[7]Ibid.

TABLE A-1 Overview of Approach to SBIR-STTR Programs Assessment

SBIR/STTR Assessment Parameters →	Quality of Research	Commercialization of SBIR-/STTR-Funded Research/Economic and Non-economic Benefits	Small Business Innovation/ Growth	Use of Small Businesses to Advance Agency Missions
Questions	How does the quality of SBIR-/STTR-funded research compare with that of other government funded R&D?	How effectively does SBIR/STTR support the commercialization of innovative technologies? What non-economic benefits can be identified?	How to broaden participation and expand the base of small innovative firms	How to increase agency support for commercializ-able technologies while continuing to support high-risk research
Measures	Peer-review scores, publication counts, citation analysis	Sales, follow-up funding, other commercial activities	Patent counts and other intellectual property/employ-ment growth, number of new technology firms	Innovative products resulting from SBIR/STTR work
Tools	Case studies, agency program studies, study of repeat winners, bibliometric analysis	Phase II surveys, program manager discussions, case studies, study of repeat winners	Phase I and Phase II surveys, case studies, study of repeat winners	Program manager surveys, case studies, agency program studies, study of repeat winners
Key Research Challenges	Difficulty of measuring quality and of identifying proper reference group	Skew of returns; significant interagency and inter-industry differences	Measures of actual success and failure at the project and firm levels; relationship of federal and state programs in this context	Major interagency differences in use of SBIR/STTR to meet agency missions

NOTE: Supplementary tools may be developed and used as needed. In addition, since publication of the methodology report, this committee has determined that data on outcomes from Phase I awards are of limited relevance.
SOURCE: National Research Council, *An Assessment of the Small Business Innovation Research Program: Project Methodology*, Washington, DC: The National Academies Press, 2004, Table 1, p. 3.

- **Case studies.** In-depth case studies of 11 STTR recipients at the five study agencies. These companies were geographically and demographically diverse and were at different stages of the company lifecycle.
- **Workshops.** A workshop in 2015 on STTR to allow stakeholders, agency staff, and academic experts to provide insights into the programs' operations, as well as to identify questions that should be addressed.
- **Analysis of agency data.** The agencies provided a range of datasets covering various aspects of agency STTR activities.
- **Open-ended responses from STTR recipients.** For the first time, survey responses included textual answers to provide a deeper view into certain questions. More than 500 responses were generated.
- **Agency meetings.** We discussed program operations with staff at all five study agencies, drawing out information both about the program and the challenges that they faced.
- **Literature review.** Since the start of our research in this area, a number of papers have been published addressing various aspects of the SBIR-STTR programs. In addition, other organizations—such as the Government Accountability Office (GAO)—have reviewed particular parts of the SBIR-STTR programs. Where useful, references to these works have been included in the course of this analysis.

Taken together with our deliberations and the expertise brought to bear by individual committee members, these tools provide the primary inputs into the analysis. For both the SBIR reports and for the current study, multiple research methodologies feed into every finding and recommendation. No finding or recommendation rested solely on data and analysis from the survey; conversely, survey data were used to support analysis throughout the report.

COMMERCIALIZATION METRICS AND DATA COLLECTION

Recent congressional interest in the SBIR-STTR programs has to a considerable extent focused on bringing innovative technologies to market. This enhanced attention to the economic return from public investments made in small business innovation is understandable. In its 2008 report on the SBIR program,[8] the committee charged with the first-round assessment held that a binary metric of commercialization was insufficient. It noted that the scale of commercialization is also important and that there are other important milestones both before and after the first dollar in sales that should be included in an appropriate approach to measuring commercialization.

[8]National Research Council, *An Assessment of the SBIR Program*, Washington, DC: The National Academies Press, 2008.

Challenges in Tracking Commercialization

Despite substantial efforts by the agencies, significant challenges remain in tracking commercialization outcomes for the STTR program. These include the following:

- **Data limitations.** Data tracking at the agencies varies widely in scale and scope. DoD and DoE utilize a similar web-based system, NSF uses a telephone-based approach, and NIH and NASA are developing their tracking programs.
- **Linear linkages.** Tracking efforts usually seek to link a specific project to a specific outcome. Separating the contributions of one project is difficult for many companies, given that multiple projects typically contribute to both anticipated and unanticipated outcomes.
- **Lags in commercialization.** Data from the extensive DoD commercialization database suggest that most projects take at least 2 years to reach the market *after the end of the Phase II award.* They do not generate peak revenue for several years after this. Therefore, efforts to measure program productivity must account for these significant lags.
- **Attribution problems.** Commercialization is often the result of several awards, not just one, as well as other factors, so attributing company-level success to specific awards is challenging at best.

Why New Data Sources Are Needed

Congress often seeks evidence about the effectiveness of programs or indeed about whether they work at all. This interest has in the past helped to drive the development of tools such as the Company Commercialization Report (CCR) at DoD, which captures the quantitative commercialization results of companies' previous Phase II projects. However, in the long term the importance of tracking may rest more in its use to support program management. By carefully analyzing outcomes and CCR's associated program variables, program managers will be able to manage their STTR portfolios more successfully.

In this regard, the STTR program can benefit from access to the survey data. The survey work provides quantitative data necessary to provide an evidence-driven assessment and, at the same time, allows management to focus on specific questions of interest, in this case related to operations of the STTR program itself.

SURVEY ANALYSIS

Traditional modes of assessing the SBIR-STTR programs include case studies, meetings, and other qualitative methods of assessment. These remain important components of the overall methodology, and a chapter in the current report is devoted to lessons drawn from case studies. However, qualitative assessment alone is insufficient.

2011-2014 Survey

The 2011-2014 Survey offers some significant advantages over other data sources. Specifically, it:

- provides a rich source of textual information in response to open-ended questions;
- probes more deeply into company demographics and agency processes;
- for the first time addresses principal investigators (PIs), not just company business officials;
- allows comparisons with previous data-collection exercises; and
- addresses other Congressional objectives for the program beyond commercialization.

For these and other reasons, we determined that a survey would be the most appropriate mechanism for developing quantitative approaches to the analysis of the STTR programs. At the same time, however, we are fully cognizant of the limitations of survey research in this case. Box A-1 describes a number of areas where caution is required when reviewing results.

This report in part addresses the need for caution by publishing the number of responses for each question and indeed each subgroup. As noted later in this discussion, the use of a control group was found to be infeasible.

Non-respondent Bias

The committee is aware that it is good practice where feasible to ascertain the extent and direction of non-respondent bias. We also acknowledge the likelihood that data from the survey may be affected by the undoubted survey deployment bias toward surviving firms.

Very limited information is available about SBIR/STTR award recipients: company name, location, and contact information for the PI and the company point of contact, agency name, and date of award (data on woman and minority ownership are not considered reliable). No detailed data are available on applicants who did not win awards. It is therefore not feasible to undertake detailed analysis of non-respondents, but the possibility exists that they would present a different profile than would respondents.

Non-respondent bias may of course work in more than one direction. Unsuccessful firms go out of business, but successful firms are often acquired by larger firms. As they are absorbed, staff are dissipated and units rearranged until PIs from these successful firms are also often unreachable. This is an especially significant instance of non-response bias in this case, as the well-known skew in outcomes for high-tech firms suggests that some of the most successful firms and projects are beyond the reach of the survey, and outcomes from these firms may account for a substantial share of overall outcomes from the program.

These inevitable gaps among both successful and unsuccessful firms are compounded by the substantial amount of movement by PIs independent of firm outcomes. PIs move to new firms, move to academia, retire, or in some cases die. In almost all cases, their previous contact information becomes unusable. Although in theory it is possible to track PIs to a new job or into retirement, in practice and given the resources available, the committee did not consider this to be an appropriate use of limited funding.

Finally, in its recent study of the SBIR program at DoD,[9] the committee compared outcomes drawn from the Academies survey and the CCR database and found that, where there was overlap in the questions, outcomes were approximately similar even though the DoD database is constructed using a completely different methodology and is mandatory for all firms participating in the SBIR-STTR programs. Although equivalent cross-checks are not available for the other agencies, the comparison with CCR data does provide a direct cross-check for one-half of all SBIR/STTR awards made and also suggests that the Academies survey methodology generates results that can be extended with some confidence to the other study agencies.

DEPLOYMENT OF THE 2011-2014 ACADEMIES PHASE II SURVEY

The Academies contracted with Grunwald Associates LLC to administer surveys to DoD, NASA, and NSF Phase II award recipients in fall 2011 and to NIH and DoE recipients in 2014. Delays in contracting with NIH and DoE resulted in the two-track deployment noted above. The Academies' 2011-2014 Survey is built closely on the previous 2005 Survey, but it is also adapted to draw on lessons learned and includes some important changes discussed in detail below. A subgroup of this committee with particular expertise in survey methodology also reviewed the survey and incorporated current best practices.[10]

---

[9]National Academies, *SBIR at the Department of Defense*, Washington, DC: The National Academies Press, 2014.

[10]Delays at NIH and DoE in contracting, combined with the need to complete work contracted with DoD, NSF, and NASA led us to proceed with the survey at the remaining three agencies first, in 2011, followed by the NIH-DoE survey in 2014.

BOX A-1
Multiple Sources of Bias in Survey Response[a]

Large innovation surveys involve multiple sources of potential bias that can skew the results in different directions. Some potential survey biases are noted below.

- Successful and more recently funded companies are more likely to respond. Research by Link and Scott demonstrates that the probability of obtaining research project information by survey decreases for less recently funded projects and increases the greater the award amount.[b] Winners from more distant years are difficult to reach: small businesses regularly cease operations, are acquired, merge, or lose staff with knowledge of SBIR/STTR awards. This may skew commercialization results downward, because more recent awards will be less likely to have completed the commercialization phase.
- Success is self-reported. Self-reporting can be a source of bias, although the dimensions and direction of that bias are not necessarily clear. In any case, policy analysis has a long history of relying on self-reported performance measures to represent market-based performance measures. Participants in such retrospective analyses are believed to be able to consider a broader set of allocation options, thus making the evaluation more realistic than data based on third-party observation.[c] In short, company founders and/or PIs are in many cases simply the best source of information available.
- Survey sampled projects from PIs with multiple awards. Projects from PIs with large numbers of awards were underrepresented in the sample, because PIs could not be expected to complete a questionnaire for each of numerous awards over a 10-year time frame.
- Failed companies are difficult to contact. Survey experts point to an "asymmetry" in the survey's ability to include failed companies for follow-up surveys in cases where the companies no longer exist.[d] It is worth noting that one cannot necessarily infer that the SBIR/STTR project failed; what is known is only that the company no longer exists.
- Not all successful projects are captured. For similar reasons, the survey could not include ongoing results from successful projects in companies that merged or were acquired before and/or after commercialization of the project's technology. This is the outcome for many successful companies in this sector.
- Some companies are unwilling to fully acknowledge SBIR/STTR contribution to project success. Some companies may be unwilling to acknowledge that they received important benefits from participating in

public programs for a variety of reasons. For example, some may understandably attribute success exclusively to their own efforts.

- Commercialization lag. Although the 2005 Survey broke new ground in data collection, the amount of sales made—and indeed the number of projects that generated sales—are inevitably undercounted in a snapshot survey taken at a single point in time. On the basis of successive data sets collected from SBIR/STTR award recipients, it is clear that total sales from all responding projects will be considerably greater than can be captured in a single survey.[e] This underscores the importance of follow-on research based on the now-established survey methodology. Figure Box A-1 illustrates this impact in practice at DoD: projects from fiscal year 2006 onward have not yet completed commercialization as of August 2013.

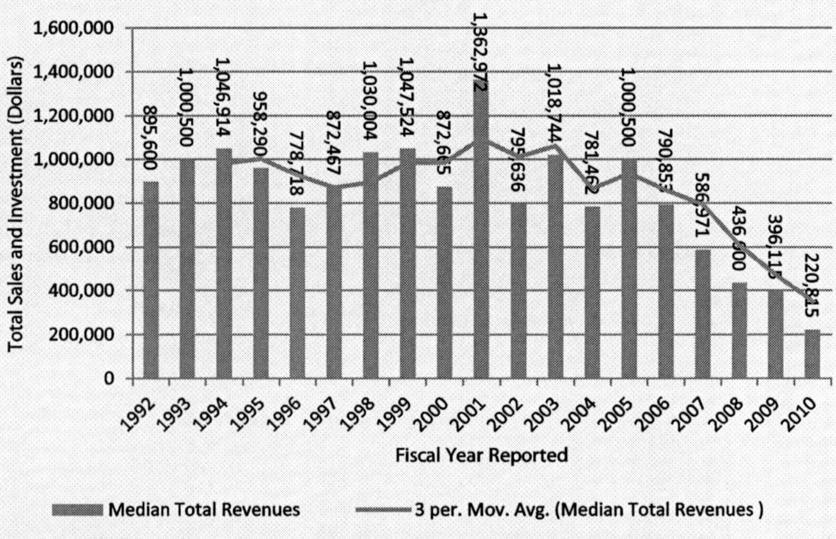

FIGURE Box A-1 The impact of commercialization lag.
SOURCE: DoD Company Commercialization Database.

[a] The limitations described here are drawn from the methodology outlined for the previous survey in National Research Council, *An Assessment of the SBIR Program at the Department of Defense*, Washington, DC: The National Academies Press, 2009.

[b] Albert N. Link and John T. Scott, *Evaluating Public Research Institutions: The U.S. Advanced Technology Program's Intramural Research Initiative*, London: Routledge, 2005.

[c] Although economic theory is formulated on what is called "revealed preferences," meaning that individuals and companies reveal how they value scarce resources by how they allocate those resources within a market framework, quite often expressed preferences are a better source of information, especially from an evaluation perspective. Strict adherence to a revealed preference paradigm could lead to misguided policy conclusions because the paradigm assumes that all policy choices are known and understood at the time that an individual or company reveals its preferences and that all relevant markets for such preferences are operational. See Gregory G. Dess and Donald

W. Beard, "Dimensions of organizational task environments," *Administrative Science Quarterly*, 29: 52-73, 1984; Albert N. Link and John T. Scott, *Public Accountability: Evaluating Technology-Based Institutions*, Norwell, MA: Kluwer Academic Publishers, 1998.
[d] Albert N. Link and John T. Scott, *Evaluating Public Research Institutions.*
[e] Data from the National Research Council assessment of the SBIR program at NIH indicate that a subsequent survey taken 2 years later would reveal substantial increases in both the percentage of companies reaching the market and the amount of sales per project. See National Research Council, *An Assessment of the SBIR Program at the National Institutes of Health*, Washington, DC: The National Academies Press, 2009.

The primary objectives of the 2011-2014 Survey are to

- provide an update of the program "snapshot" taken in 2005, maximizing the opportunity to identify trends within the program;
- probe more deeply into program processes, with the help of expanded feedback from participants and better understanding of program demographics; and
- reduce costs and shrink the time required by combining three 2005 questionnaires—for the company, Phase I, and Phase II awards, respectively—into a single survey questionnaire.

The survey was therefore designed to collect the maximum amount of data, consistent with our commitment to minimizing the burden on individual respondents.

In light of these competing considerations, the committee determined that it would be more useful and effective to administer the survey to PIs—the lead researcher on each project—rather than to the registered company point of contact (POC), who in many cases would be an administrator rather than a researcher. This decision was reinforced by difficulties in accessing current POC information. Key areas of overlap between the 2005 and 2014 surveys are captured in Table A-2.

Starting Date and Coverage

The 2011-2014 Survey included awards made from fiscal year (FY)1998-2007 for DoD, DoE, and NSF and for FY2001 to FY2010 inclusive for NIH and DoE. This end date allowed for completion of Phase II-awarded projects (which nominally fund 2 years of research) and provided a further 2 years for commercialization. This time frame was consistent with the previous survey, administered in 2005, which surveyed awards from FY1992 to FY2001. It was also consistent with a previous GAO study, which in 1991 surveyed awards made through 1987.

The aim in setting the overall time frame at 10 years was to reduce the impact of difficulties in generating information about older awards because some companies and PIs may no longer be in place and memories fade over time.

TABLE A-2 Similarities and Differences: 2005 and 2014 Surveys

Item	2005 Survey	2014 Survey
Respondent selection		
Focus on Phase II winners	✓	✓
All qualifying awards		✓
PIs		✓
POCs	✓	
Max number of questionnaires per respondent	<20	2
Distribution		
Mail	✓	No
Email	✓	✓
Telephone follow-up	✓	✓
Questionnaire		
Company demographics	Identical	Identical
Commercialization outcomes	Identical	Identical
IP outcomes	Identical	Identical
Women and minority participation	✓	✓
Additional detail on minorities		✓
Additional detail on PIs		✓
New section on agency staff activities		✓
New section on company recommendations for SBIR/STTR		✓
New section on STTR		✓
New section capturing open-ended responses		✓

Determining the Survey Population

Following the precedent set by both the original GAO study and the first round of Academies analysis, we differentiate between the total population of STTR recipients, the preliminary survey target population, and the effective population for this study, which is the population of respondents that were reachable.

Initial Filters for Potential Recipients

Determining the effective study population required the following steps:

- acquisition of data from the five study agencies covering record-level lists of award recipients during the relevant fiscal years;

- elimination of records that did not fit the protocol agreed upon by the committee—namely, a maximum of two questionnaires per PI (in cases where PIs received more than two awards). In these cases, awards were selected first by program (STTR, then SBIR), then by agency (in order: NSF, NASA, and DoD for 2011 and DoE and NIH for 2014), then by year (oldest first), and finally by random number; and
- elimination of records for which there were significant missing data.

This process of excluding awards either because they did not fit the selection profile approved by the committee or because the agencies did not provide sufficient or current contact information reduced the total STTR award list for the five agencies from 1,501 awards to a preliminary survey population of 1,400 awards.

Secondary Filters to Identify Recipients with Active Contact Information

This nominal population still included many potential respondents whose contact information was formally complete in the agency records but who were no longer associated with the contact information provided and hence effectively unreachable. This is not surprising given that small businesses experience considerable turnover in personnel and that the survey reaches back to awards made in FY1998. Recipients may have switched companies, the company may have ceased to exist or been acquired, or telephone and email contacts may have changed, for example. Consequently, we utilized two further filters to help identify the effective survey population.

- First, contacts for which the email address bounced twice were eliminated. Because the survey was delivered via email, the absence of a working email address disqualified the recipient. This eliminated approximately 20 percent of the preliminary population.
- Second, email addresses that did not officially "bounce" (i.e., return to sender) may still in fact not be active. Some email systems are configured to delete unrecognized email without sending a reply; in other cases, email addresses are inactive but not deleted. So a non-bouncing email address did not equal a contactable PI. Accordingly, Grunwald Associates made efforts to contact by telephone all non-respondents at the five agencies. Up to two calls were made, and outcomes from the telephone calls were used to further filter non-contactable PIs. Thirty seven percent of the preliminary population was non-contactable by telephone.[11]

[11]This percentage includes only those individuals whose telephone contact information was clearly no longer current, for example, the phone number was invalid, the company was out of business, or the PI no longer worked at the company.

Deployment

The 2011 Survey opened in fall 2011 and the 2014 Survey in winter 2014. Both were deployed by email, with voice follow-up support. Up to four emails were sent to the effective population (emails discontinued once responses were received). In addition, two voice mails were delivered to non-respondents between the second and third and between the third and fourth rounds of email. In total, up to six efforts were made to reach each questionnaire recipient. The surveys were open for 11 and 18 weeks, respectively.

Response Rates

Standard procedures were followed to conduct the survey. These data collection procedures were designed to increase response to the extent possible within the constraints of a voluntary survey and the survey budget. The population surveyed is a difficult one to contact and obtain responses from, as evidence from the literature shows. Under these circumstances, the inability to contact and obtain responses always raises questions about potential bias of the estimates that cannot be quantified without substantial extra efforts that would require resources beyond those available for this work.

Table A-3 shows the response rates for STTR at the five agencies, based on both the preliminary study population prior to adjustment and the effective study population after all adjustments.

Effort at Comparison Group Analysis

Several readers of the reports in the first-round analysis of the SBIR-STTR programs suggested the inclusion of comparison groups in the analysis.

TABLE A-3 2011-2014 STTR Survey Response Rates

	Total
Total Awards	1,501
Excluded from survey population	101
Preliminary target population	1,400
Not contactable	807
Bad emails	266
Bad phone	518
Opt outs	23
Effective survey population	593
Completed surveys	292
Success rate (preliminary population)	20.9
Success rate (effective population)	49.2

SOURCE: 2011-2014 Survey.

We concurred that this should be attempted. There is no simple and easy way to acquire a comparison group for Phase II SBIR/STTR awardees. These are technology-based companies at an early stage of company development, which have the demonstrated capacity to undertake challenging technical research *and* to provide evidence that they are potentially successful commercializers. Given that the operations of the SBIR-STTR programs are defined in legislation and limited by the Small Business Administration (SBA) Policy Guidance, randomly assigned control groups were not a possible alternative. Efforts to identify a pool of SBIR/STTR-like companies were made by contacting the most likely sources—Dunn and Bradstreet and Hoovers—but these efforts were not successful, because sufficiently detailed and structured information about companies was not available.

In response, the committee sought to develop a comparison group from among Phase I awardees that had not received a Phase II award from the three surveyed agencies (DoD, NSF, and NASA) during the award period covered by the 2011 Survey (FY1998-2007). After considerable review, however, we concluded that the Phase I-only group was not appropriate for use as a statistical comparison group, because the latter was not deemed to be a sufficiently independent control group.

Responses and Respondents

Table A-4 shows STTR responses by year of award. The survey primarily reached companies that were still in business—overall, 83 percent of respondents.[12]

[12]2011-2014 Survey, Question 4A.

TABLE A-4 STTR Responses by Year of Award (Percent Distribution)

Fiscal Year of Award	STTR
1998	
1999	0.7
2000	0.7
2001	4.8
2002	6.5
2003	5.5
2004	7.2
2005	14.4
2006	14.4
2007	19.2
2008	7.5
2009	7.2
2010	12.0
Total	100.0
BASE: ALL RESPONDENTS	292

SOURCE: 2011-2014 Survey.

Appendix B

Major Changes to the SBIR and STTR Programs Resulting from the 2011 SBIR Reauthorization Act, P.L. 112-81, December 2011

1) **The SBIR program received an increased share of federal agencies' extramural budget:** [1]

    a. Congress increased the SBIR/STTR share from 2.5 percent to 2.6 percent in FY2012 and by 0.1 percent per year through FY2016 and by 0.2 percent in FY2017, when the share would be 3.2 percent.

2) **STTR's share of the overall combined program was increased:** [2]

    a. The STTR share was 0.3 percent from FY2004 through FY2011, 0.35 percent in FY2012 and 2013, 0.40 percent in FY 2014 and 2015, and 0.45 percent in 2016 and thereafter.

3) **Award levels were increased:**[3]

    a. The existing limit of $100,000 for Phase I SBIR and STTR awards was increased to $150,000.
    b. The existing limit of $750,000 for Phase II SBIR and STTR awards was increased to $1,000,000.
    c. These limits were also for the first time indexed to inflation.

---

[1]U.S. Congress, P.L. 112-81, Sec. 5102 (a)(1)(a).
[2]Sec. 5102(b).
[3]Sec. 5103.

4) **Agency flexibility to issue larger awards was curtailed:**[4]

   a. Awards may no longer exceed 150 percent of guidelines (i.e., $1.5 million for Phase II) without a specific waiver from the SBA Administrator.
   b. The waiver can apply only to a specific topic, not to the agency as a whole. The agency must meet specific criteria and must show in its application that these criteria have been met before a waiver can be issued.
   c. For every award under a waiver, agencies must maintain additional information about the recipient, including the extent to which they are owned or funded by venture capital or hedge fund investors.

5) **Agencies are permitted to utilize awards from other agencies:**[5]

   a. Agencies gained the ability to adopt Phase I awards from other agencies for Phase II funding; however, senior agency staff must certify that this is appropriate.
   b. Similarly, the legislation now permits between-phase crossovers between SBIR and STTR.

6) **Phase II invitations were eliminated for SBIR:** [6]

   a. The requirement that a company be invited by the agency before it could propose work for Phase II is now eliminated.

7) **Pilot programs to skip Phase I were established:**[7]

   a. The legislation allows NIH, DoD, and the Department of Education to undertake pilot programs in this area. Discussions with agency staff indicate that for now DoD does not expect to utilize this new flexibility.

8) **For SBIR. limited participation by previously excluded firms with majority venture capital or hedge fund ownership is now permitted (although subsidiaries of large operational companies are still excluded):** [8]

---

[4]Sec. 5103.
[5]Sec. 5104.
[6]Sec. 5105.
[7]Sec. 5106.
[8]Sec. 5107.

a. NIH, NSF, and DoE are permitted to award up to 25 percent of their program funding to such companies.
b. Other agencies are limited to 15 percent.
c. For each award to such an entity, the Agency or component head must certify that this award is in the public interest based on criteria laid out in Sec. 5107(A)(dd)(2).
d. Access to venture capital or hedge fund support may not be used as an award selection criterion by agencies.
e. Special "affiliation" rules are provided for venture capital- and hedge fund-owned companies:

   i. Portfolio companies partially owned by venture firms or hedge funds are not deemed to be "affiliated" for purposes of determining whether an applicant meets size limitations, unless they are wholly owned or the owning company has a majority of board seats on the portfolio company.

9) Explicit procurement preference were given for SBIR and STTR projects:[9]

a. The legislation states that agencies *and prime contractors* (emphasis added) must give preference to SBIR and STTR projects where practicable. However, there are no explicit targets included in the legislation.

10) Sequential Phase II awards were permitted: [10]

a. The legislation now explicitly permits agencies to award one additional Phase II award after the first Phase II has been completed.
b. The language implies that the provision of more than one sequential Phase II is prohibited.

11) **Commercialization support was expanded:**[11]

a. Agencies are permitted to spend up to $5,000 per year per award on support for commercialization activities.
b. Individual firms can now request up to $5,000 per year *in addition to their SBIR or STTR award* (emphasis added) to pay for commercialization activities from agency-approved vendors.

---

[9]Sec. 5108.
[10]Sec. 5111.
[11]Sec. 5121.

12) **The commercialization readiness pilot at DoD was converted to a permanent program—the Commercialization Readiness Program (CRP).** Details include in particular the following:[12]

   a. An SBIR Phase III insertion plan is now required for all DoD acquisition programs with a value of $100 million or more.
   b. SBIR/STTR Phase III reporting is now required from the prime contractor for all such contracts.
   c. The Secretary of Defense (SecDef) is now required to set goals for the inclusion of SBIR/STTR Phase II projects in programs of record and fielded systems and must report on related plans and outcomes to the SBA Administrator.
   d. The legislation explicitly requires the SecDef to develop incentives toward this purpose and to report on the incentives and their implementation.

13) **CRP may be expanded to other agencies:**[13]

   a. Other agencies may spend up to 10 percent of their SBIR-STTR program funds on commercialization programs.
   b. CRP awards may be up to three times the maximum size of Phase II awards.
   c. CRP authority expires after FY2017.

14) **Phase 0 pilot partnership program at NIH was enabled:**[14]

   a. NIH is permitted to use $5 million to establish a Phase 0 pilot program.
   b. The funding must go to universities or other research institutions that participate in the NIH STTR program.
   c. These institutions must then use the funding for Phase 0 projects for individual researchers.

15) **Data collection and reporting were enhanced:**[15]

   a. Overall, the legislation calls for substantially increased data collection for individual recipients and for much more detailed reporting from agencies to SBA and to Congress.

---

[12]Sec. 5122.
[13]Sec. 5123.
[14]Sec. 5127.
[15]Especially Sec. 5132, Sec. 5133, Sec. 5138, and Sec. 5161, but specific requirements are found throughout the legislation.

b. Specific areas for improved reporting include:

   i. Participation of (and outreach toward) woman- and minority-owned firms and the participation of woman and minority principal investigators;
   ii. Phase III take-up (from both agencies and prime contractors);
   iii. Participation of venture capital- and hedge fund-owned firms;
   iv. Appeals and noncompliance actions taken by SBA;
   v. Sharing of data between agencies electronically;
   vi. Extra-large awards;
   vii. SBIR and STTR project outcomes (from participants);
   viii. University connections (especially for STTR projects);
   ix. Relations with the FAST state-level programs;
   x. Use of administrative funding for SBIR;
   xi. Development of program effectiveness metrics at each agency; and
   xii. SBIR activities related to Executive Order 1339 in support of manufacturing.

c. SBA is charged with developing a unified database to cover all SBIR and STTR awards at all agencies, as well as company information and certifications.[16]

16) Funding was provided for a pilot program to cover administrative, oversight, and contract processing costs:[17]

a. Agencies are limited to spending 3 percent of their SBIR funding on this pilot.
b. The pilot is initially designated to last for 3 fiscal years following enactment.
c. Part of the funding must be spent on outreach in low-award states.

17) Minimum commercialization rates for participating companies are required:[18]

a. Agencies must establish appropriate commercialization metrics and benchmarks for participating companies, for both Phase I and Phase II (subject to SBA Administrator approval).
b. Failure to meet those benchmarks must result in 1-year exclusion for that company from the agency's SBIR and STTR programs.

---

[16]Sec. 5135.
[17]Sec. 5141.
[18]Sec. 5165.

Appendix C

National Academy of Sciences, Engineering, and Medicine 2014 SBIR/STTR Survey

Introduction

Welcome to the National Academies SBIR/STTR Survey. Thank you for participating. This survey seeks responses related to the Phase II project entitled [insert project title], funded by [insert agency name], at [insert company name]. Funding was awarded in [insert FY].

Note: If you need to revisit the survey before finally completing it, you can return at the point you left off by clicking on the survey link in your email.

Finally, please use the navigational buttons within the survey. The back and forward buttons on your browser will not work.

Privacy and Confidentiality Policy

Responses to this survey will be held in confidence by the survey team. No identifiable information will be provided to other Academy staff or to the Public Access File which provides researchers with access to project data.

In order to implement this commitment, the following steps have been taken, covering three areas:

a) Data in the published report
b) Management of raw data files
c) Additional review of textual (open-ended) responses

a) Data in the published report.

All data except for text responses will be presented only in aggregated form in the report; no individually identifiable cells will be published.

b) Managing raw data.

In order to provide researchers with access while meeting the confidentiality commitment, the following steps will be taken by the Contractor prior to providing an expurgated data set to the Academy for inclusion in the Public Access File:

1) Replace company name with a new company ID
2) Replace PI name with a new PI ID
3) Delete the following fields:

    a. Agency record ID
    b. Company address except for State field
    c. Project title
    d. Project abstract
    e. Flag for woman owned business
    f. Flag for minority owned business

The raw (unexpurgated) data set will be retained by the Contractor for two years after publication of the report. All copies of the raw data will then be destroyed. The expurgated data set will be retained indefinitely in the Public Access File related to the project.

c) Review of textual responses.

Two independent reviewers will analyze open ended responses with a view to redacting material that could provide clues as to the identity of the respondent prior to their inclusion in the Public Access File. In particular, this review will redact all company names, product names, and PI or other company official names, as well as other potential identity clues.

Do you approve the privacy and confidentiality policy as shown above?
[Yes/No. If no, jump to page 55.]

This information is required only to determine your current status, and to ensure that we have accurate contact information. Your information will be strictly private and will not be shared with any private entity or government agency; aggregated data will be shared in a published report.

1. For the project referenced above, were you (during the time period covered by this award) …*

Select all that apply.

a. A Principal Investigator (PI) on this project
b. The CEO
c. A company founder
d. Senior researcher (other than PI)
e. Not CEO but a senior executive with the company identified above
f. None of the above (exit questionnaire)

Part 1. Information About You.

2. Please verify or correct the following information about yourself. Please indicate any corrections in the boxes provided. If all this information is accurate, click "Next to continue.

First name: [Text box]
Last name: [Text box]
Current email address: [Text box]
Current work telephone number (for follow up questions if necessary): [Text box]

Part 2. Company Information Section

3. Have you already completed a questionnaire about another SBIR or STTR project for this National Academy survey related to [insert company name]?*

[Yes/No. If yes, skip to Part 3: PI/Senior Executive Information]

4. Is [insert company name] still in business?

[Yes/No]

5. Thinking about the number of founders of the company, what was…?

Min = 0 Max = 20 Must be numeric

a. The total number of founders [number box]
b. The number of other companies started by one or more of the founders (before starting this one) [number box]
c. The number of founders who have a business background [number box]
d. The number of founders who have an academic background [number box]
e. The number of founders with previous experience as company founders [number box]

6. What was the most recent employment of the company founders prior to founding the company?

   Select all that apply.
   a. Other private company
   b. Government
   c. Research institution
   d. FFRDCs or National Labs
   e. Other

7. Was the company founded because of the SBIR/STTR program?

   Yes
   In part
   No

8. What was the company's total revenue for the most recent fiscal year?

   $0
   Under $100,000
   $100,000-499,999
   $500,000-999,999
   $1,000,000-4,999,999
   $5,000,000-19,999,999
   $20,000,000-99,999,999
   $100,000,000 or more

9. What percentage of the company's revenues during its most recent completed fiscal year was Federal SBIR/STTR funding (Phase I and/or Phase II)?

   0%
   1-10%
   11-25%
   26-50%
   51-75%
   76-99%
   100%

10. What percentage of the company's total R&D effort (man-hours of scientists and engineers) was for SBIR/STTR activities during the most recent fiscal year?

    0%
    1-10%
    11-25%
    26-50%
    51-75%
    76-100%

11. Which if any of the following has the firm experienced since your first SBIR/STTR award?

    Select all that apply.
    Made an initial public offering
    Established one or more spin off companies
    Been acquired by/merged with another firm
    Planning to make an initial public offering in the next two years
    Entered into strategic partnership with major industry player
    None of the above

12. How many patents have resulted, at least in part, from the company's SBIR/STTR awards?

    Min = 0 Max = 999 Must be numeric
    Whole numbers only
    Positive numbers only

    [number box]

13. Does the company have one or more full time staff for marketing or business development?
    [Yes/No]

14. Number of company employees (including all affiliates):

    Min = 0 Max = 99999 Must be numeric
    Whole numbers only
    Positive numbers only

    a. At the time of the award in [pipe in award year] [Number box]
    b. Currently [Number box]

15. What was the ownership status of the company at the time of the award?

    Select all that apply.
    a. Woman-owned
    b. Minority-owned
    c. Neither of the above

    If the answer is "Minority-owned," please indicate the ethnic minority group[s] that company owners [at the time of the award] belonged to.

    Select all that apply.
    Asian-Indian
    Asian-Pacific
    Black
    Hispanic
    Native American
    Other [Text box]

Part 3. PI/Senior Executive Information

16. The Principal Investigator for this [SBIR/STTR] Award was a …

    Select all that apply.
    a. Woman
    b. Minority
    c. Neither of the above

    If the answer is "Minority," please indicate the ethnic minority group[s] the Principal Investigator for this award belongs to.

    Select all that apply.
    Asian-Indian
    Asian-Pacific
    Black
    Hispanic
    Native American
    Other [Text box]

17. At the time of the award, the age of the leading PI was…

    [Under 25, 25-29, 30-34, 35-39, 40-44, 45-49, 50-54, 55-59, 60-64, 65+]

18. What was the immigration status of the PI at the time of the award?

American-born US citizen
Naturalized US citizen
US Green card
H1 visa
Other [Text box]

19. What is the current status of the project funded by the referenced award?

Select the one best answer.

a) Project has not yet completed SBIR/STTR funded research.
b) Efforts at this company have been discontinued. No sales or additional funding resulted from this project.
c) Efforts at this company have been discontinued. The project did result in sales, licensing of technology, or additional funding.
d) Project is continuing post-award technology development.
e) Commercialization is underway.
f) Products/Processes/Services are in use by target population/customer/consumers.
g) Products/Processes/Services are in use by population/customer/consumers not anticipated at the time of the award (for example, in a different industry).

20. If the answer is either b) or c), did the reasons for discontinuing this project include any of the following?

Select one of the reasons as the Primary Reason. Select all that apply as Other contributing reasons.

Another firm got to the market before us
Level of technical risk too high
Principal Investigator left
Technical failure or difficulties
Inadequate sales capability
Project goal was achieved (e.g. prototype delivered for federal agency use)
Licensed to another company
Market demand too small
Company shifted priorities
Other (Please specify in comments box below)

Comments
[Text box]

Part 4. Project status information

21. Please select the technology sector or sectors that most closely fit(s) the work of the SBIR/STTR project.

Select all that apply.

Aerospace and Defense
- Aerospace
- Defense-specific products and services

Energy and the environment
- Renewable energy production (solar, wind, geothermal, bio-energy, wave)
- Energy storage and distribution
- Energy efficiency
- Other energy or environmental products and services

Engineering
- Engineering services
- Scientific instruments and measuring equipment
- Robotics
- Sensors
- Other engineering

Information technology
- Computers and peripheral equipment
- Telecommunications equipment and services
- Business and productivity software
- Data processing and database software and services
- Media products (including web-, print- and wireless-delivered content)
- Other IT

Materials
- Materials (including nanotechnology for materials)

Medical technologies
- Pharmaceuticals
- Medical devices
- Biotechnology (including therapeutic, diagnostic, combination)
- Health IT (including mobile, big data, training modules)
- Research tools
- Other medical products and services

Other (please specify) [Text box]

22. Did you experience a gap between the end of Phase I and the start of Phase II for this award?

[Yes/No. If no, skip to question 25]

23. During the funding gap between Phase I and Phase II for this award, which of the following occurred?

    Select all answers that apply.

    a. Stopped work on this project during funding gap.
    b. Continued work at reduced pace during funding gap.
    c. Continued work at pace equal to or greater than Phase I pace during funding gap.
    d. Received gap funding between Phase I and Phase II.
    e. Company ceased all operations during funding gap
    f. Other (specify) [Text box]

24. In your opinion, in the absence of this SBIR/STTR award, would the company have undertaken this project?

    a. Definitely yes [Answer questions 25-27.]
    b. Probably yes [Answer questions 25-27.]
    c. Uncertain
    d. Probably not
    e. Definitely not

25. If you had undertaken this project in the absence of SBIR/STTR, this project would have been …
    a. Broader in scope
    b. Similar in scope
    c. Narrower in scope

26. In the absence of SBIR/STTR funding… (Please provide your best estimate of the impact)

    a. …how long would the start of this project have been delayed? [text box] months
    b. …the expected duration/time to completion would have been…
       1) longer
       2) the same
       3) shorter
    c. …in achieving similar goals and milestones, the project would be…
       1) ahead
       2) the same place
       3) behind

27. Did this award require matching funds or other types of cost sharing in the Phase II Proposal?
    [Yes/No. If No, skip questions 28-39.]

28. Matching or co-investment funding proposed for Phase II was received from ...

    Select all that apply.
    Non-SBIR/STTR federal funds
    a. **Private investment: U.S. Sources**
        i) venture capital (VC)
        ii) U.S. angel funding or other private equity investment (not VC)
        iii) Friends and family
        iv) Strategic investors/partners
        v) Other sources
    b. **Foreign investment**
        i) Financial investors
        ii) Strategic investors/partners
    c. **Other sources**
        (1) State or local governments
        (2) Research institutions (such as colleges, universities or medical centers)
    d. **Internal sources**
        (1) Your own company (Including money you have borrowed)
        (2) Personal funds

29. How difficult was it for the company to acquire the funding needed to meet the matching funds requirements?

    a. No additional effort needed except paperwork
    b. Less than 2 weeks Full Time Equivalent (FTE) for senior company staff
    c. 2-8 weeks effort FTE for senior company staff
    d. 2-6 months of effort FTE for senior company staff
    e. More than 6 months of effort FTE for senior company staff

Part 5. Project outcomes

30. To date, what has been the total additional developmental funding for the technology developed during this project?

    None $0
    Under $100,000
    $100,000-499,999
    $500,000-999,999
    $1,000,000-4,999,999
    $5,000,000-9,999,999
    $10,000,000-19,999,999
    $20,000,000-49,999,999
    $50,000,000 or more

31. What have been the sources of additional development funding?

    Select all that apply.

    Non-SBIR/STTR federal funds

    a. **Private investment: U.S. Sources**
        i) venture capital (VC)
        ii) U.S. angel funding or other private equity investment (not VC)
        iii) Friends and family
        iv) Strategic investors/partners
        v) Other sources
    b. **Foreign investment**
        i) Financial investors
        ii) Strategic investors/partners
    c. **Other sources**
        (1) State or local governments
        (2) Research institutions (such as colleges, universities or medical centers)
    d. **Internal sources**
        (1) Your own company (Including money you have borrowed)
        (2) Personal funds

32. Has the company and/or licensee had any actual sales of products, processes, services or other sales incorporating the technology developed during this project?

Select all that apply.

a. No sales to date nor are sales expected. [Skip questions 33-39.]
b. No sales to date, but sales are expected. [Skip to question 33-39.]
c. Sales of product(s)
d. Sales of process(es)
e. Sales of services(s)
f. Other sales (e.g. rights to technology, licensing, etc.)

33. For the company and/or the licensee(s), when did the first sale occur resulting from the technology developed during [name of project]?

If multiple SBIR/STTR awards contributed to the ultimate commercial outcome, report only the share of total sales appropriate to this SBIR/STTR project.

For the company [Pulldown with choices from 1990-2014]
For any licensees [Pulldown with choices from 1990-2014]

34. For the company, what is the approximate amount of total sales dollars of product(s), process(es) or services to date resulting from the technology developed during the [name of project]?

[Pulldown with choices:
None $0
Under $100,000
$100,000-$499,999
$500,000-$999,999
$1,000,000-$4,999,999
$5,000,000-$9,999,999
$10,000,000-$19,999,999
$20,000,000-$49,999,999
$50,000,000 or more]

35. What is the approximate amount of other total sales dollars (e.g. rights to technology, sale of spin-off company, etc.) to date resulting from the technology developed during the [name of project]?

[Pulldown with choices:
None $0
Under $100,000
$100,000-$499,999
$500,000-$999,999
$1,000,000-$4,999,999
$5,000,000-$9,999,999
$10,000,000-$19,999,999
$20,000,000-$49,999,999
$50,000,000 or more]

36. To date, approximately what percent of total sales from the technology developed during this project have gone to the following customers?

Round percentages. Answers required to add to 100%.

a. Domestic private sector [Number box]
b. Export Markets [Number box]
c. Department of Defense (DoD) [Number box]
d. NASA [Number box]
e. Prime contractors for DoD [Number box]
f. Prime contractor for NASA [Number box]
g. Agency that awarded the Phase II (if not NASA or DoD) [Number box]
h. Other federal agencies [Number box]
i. State or local governments [Number box]
j. Other [Number box] (Specify below, if applicable)

If applicable please specify what "other" types of customers you have sold to as a result of this project.
[Text box]

37. Please list any significant commercial partnerships (including licensing agreements) based on the SBIR/STTR-funded technology.
[Text box]

38. Please give the number of patents, copyrights, trademarks received and articles published in scientific publications for the technology developed as **a result of** [name of project].

    Enter numbers. If none, enter 0 (zero).
    Patents [Number box]
    Copyrights [Number box]
    Trademarks [Number box]
    Published articles [Number box]

39. How many SBIR and/or STTR awards has the company received that are related to the project/technology supported by this award?

    a. Number of related Phase I awards [Text box]
    b. Number of related Phase II awards [Text box]

NIH Only

40. Does your product require FDA approval before it can be marketed?
    [Yes/No. If no, skip to question 47]

41. What is the current status of the project with regard to the FDA process?

    Process abandoned
    Preparation under way for clinical trails
    IND granted
    In Phase 1 clinical trials
    In Phase 2 clinical trials
    In Phase 3 clinical trials
    Completed clinical trials

42. What sources of funding have been employed in relation to the FDA process?

    Select all that apply.
    SBIR Phase II
    SBIR Phase IIB
    Other NIH Funding
    BARDA funding
    Internal company and personal funding
    Angel Funding
    Venture funding
    Funding from other companies
    Other (specify) [Text box]

43. For projects still in process, when approximately – assuming all goes well with clinical trials – do you anticipate completing the FDA certification process?

    [Text box]

44. What non-financial support in relation to FDA approval have you received from NIH before and during the clinical trials process?

    [Text box]

45. If applicable, how useful was this support?

    Extremely useful (5)
    4
    3
    2
    Not useful at all (1)

    Comments
    [Text box]

46. How much difference did Phase IIB funding make to the eventual outcome of the project (or its current status if research is not completed)?

    A tremendous difference (5)
    4
    3
    2
    It made no difference at all (1)

    Comments
    [Text box]

47. Was the additional funding sufficient to allow you to complete any of the following?

    Select all that apply.

    Preclinical trial preparation
    Phase 1 trials
    Phase 2 trials
    Phase 3 trials
    No/None of the above

48. What additional measures should NIH take to support companies like yours during the process?
    [Text box]

49. Many agencies offer commercialization assistance in connection with SBIR or STTR awards. Did you (or another company staff member) participate in a technical assistance related to this award that was offered by your funding agency?
    [Yes/No]

Part 6. SBIR Process and Recommendations

49. Many agencies offer commercialization assistance in connection with SBIR or STTR awards. Did you (or another company staff member) participate in a technical assistance related to this award that was offered by your funding agency?
    [Yes/No. If no, skip questions 50-73.]

    Phase I
    Phase II
    Both

50. What company provided assistance to you?

    Dawnbreaker
    LARTA
    Foresight
    Other (specify) [Text box]

51. How valuable was the commercialization assistance?

    Extremely valuable
    Very valuable
    Somewhat valuable
    Not very valuable
    Not at all valuable

52. New rules permit companies to use up to $10,000 of SBIR/STTR funding for their own marketing purposes, outside the agency program.

Would you…

Continue to use the agency's program
Use the funding for your own marketing consultant
Neither

53. In comparison to other Federal awards or Federal funding, how would you rate the process of applying for Phase II funding? Applying for SBIR/STTR Phase II funding was...

a. Much easier than applying for other Federal awards
b. Easier
c. About the same
d. More difficult
e. Much more difficult
f. Not sure, not applicable, or not familiar with other Federal awards or funding

54. How adequate was the amount of money you received through SBIR/STTR Phase II funding for the purposes you applied for? Was it...

a. More than enough
b. About the right amount
c. Not enough

55. Congress recently increased the standard limit on awards to $1 million for SBIR/STTR. Should the size of Phase II awards be increased even if that means a proportionately lower number of Phase II awards are made?

a. Yes
b. No
c. Not sure

56. Overall, would you recommend that the SBIR/STTR program be...?

    a. Expanded (with equivalent funding taken from other federal research programs that you benefit from and value)
    b. Kept at about the current level
    c. Reduced (with equivalent funding applied to other federal research programs you benefit from and value)
    d. Eliminated (with equivalent funding applied to other federal research programs you benefit from and value)

57. To what extent did the SBIR/STTR funding significantly affect long term outcomes for the company?

    a. Had a highly positive or transformative effect
    b. Had a positive effect
    c. Had no effect
    d. Had a negative effect
    e. Had a highly negative or disastrous effect

58. Can you explain these impacts in your own words?
    [Text box]

Part 7. Working with Project Managers

This section seeks information about your interactions with your agency point of contact, who for the purposes of this survey is referred to as a "Project Manager."

59. How often did you engage with your Project Manager in the course of your award?
    a. weekly
    b. monthly
    c. quarterly
    d. annually

60. How valuable was your Project Manager on a scale of 1-5, with 1 being no help and 5 being invaluable?

    Invaluable (5)
    4
    3
    2
    No help (1)

61. How knowledgeable was your Project Manager about the SBIR/STTR program. Were they able to guide you effectively through the SBIR/STTR process?

    a. Not at all knowledgeable
    b. Quite knowledgeable
    c. Somewhat knowledgeable
    d. Extremely knowledgeable

62. On a scale of 1-5, with one being least and 5 being most, how much did your project manager help during the Phase II award in the following areas: [1-5 scale for each row]

    a. Providing direct technical help
    b. Finding markets for our technology or products/services
    c. The Phase II application process
    d. Introducing us to university personnel or government labs that could contribute to the project
    e. Introducing us to other firms that could provide technical expertise

63. How closely did you work with your Project Manager as you pursued additional funding beyond Phase II?

    a. The officer provided a lot of guidance during the application process
    b. We discussed the application in detail
    c. Not much
    d. Not at all
    e. We did not apply for additional agency funding

64. How effective was the Project Manager in connecting the company to sources of Phase III funding (such as acquisition programs or venture/angel funding)?

    Very helpful
    Somewhat helpful
    Not very helpful
    Not at all helpful

65. How easy was it to reach your Project Manager when you had questions or concerns?

    Very easy
    Easy
    Hard
    Very hard

66. Was your Project Manager replaced during the course of your award?
    [Yes/No]

67. How do you see the time allocated for your Project Manager to work on your project?

    More than sufficient
    Sufficient
    Insufficient

68. Additional comments on working with your Project Manager
    [Text box]

69. Is a Federal System or Acquisition Program using the technology from this award?

    Yes (Answer question 70)
    No (Skip to question 71)

70. Please provide the name of the Federal System or Acquisition Program that is using the technology.
    [Text box]

71. This questions address any relationships between your firm's efforts on this project and any partnering Research Institution (RI) (including universities, medical centers, Federal research labs).

    Select all that apply.

    a. The PI for this project was at the time of the project an RI faculty member
    b. The PI for this project was at the time of the project an RI adjunct faculty member
    c. Faculty member(s) or adjunct faculty member(s) worked on this project in a role other than PI
    d. Graduate students worked on this project
    e. The technology for this project was licensed from an RI
    f. The technology for this project was originally developed at an RI by one of the participants in this project
    g. An RI was a subcontractor on this project
    h. None of the above [Skip questions 72-73.]

72. Which research institution (or institutions) worked with your firm on this project?
    [Text box]

73. If you worked with an FFRDC or a National Lab as part of this project, please briefly describe this aspect of the project, and add any further comments based on this aspect of the project.
    [Text box]

Part 8. STTR

74. To what extent did your STTR award change your relationship with the research institution?

    a. Substantially enhanced it
    b. Somewhat enhanced it
    c. Made no real difference
    d. Made it somewhat worse
    e. Made it substantially worse

    If you have additional comments and/or recommendations about working with a research institution in the context of SBIR/STTR, please enter them here.

75. Did you collaborate with this research institution before receiving this STTR award?
    [Yes/No]

76. Have you ever received a Small Business Innovation Research (SBIR) award?
    [Yes/No. If no, skip to question 80]

77. Have you had prior SBIR awards in which you collaborated with a research institution?
    [Yes/No]

78. From your perspective, are there significant differences between SBIR and STTR awards?
    [Yes/No. If no, skip to question 80.]

79. Please explain these differences in your own words.
    [Text box]

80. If you have received both SBIR and STTR awards, did you find that

    a. STTR is easier to manage than SBIR
    b. They are about the same
    c. STTR is harder to manage than SBIR

81. Do you think that the funding proportion that can be allocated to the research institution should be increased?

    a. Strongly agree
    b. Somewhat agree
    c. Neither agree nor disagree
    d. Somewhat disagree
    e. Strongly disagree

82. Have you tried to switch an STTR Phase I award to an SBIR Phase II award, or the other way around?
    [Yes/No]

83. Are these specific ways in which outcomes from your SBIR/STTR awards as a company have helped meet the mission of the funding agency?
    [Text box]

84. Other comments or recommendations based on your experience with the STTR program?
    [Text box]

Appendix D

List of Research Institutions Reported by STTR Survey Respondents

Research Institution	Number of Mentions
Alfred I duPont Hospital for Children	1
Argonne National Laboratory	2
Arizona State Polytechnic University	1
Baylor College of Dentistry	1
Beth Israel Deaconess Medical Center	1
Boston University	5
Boys Town National Research Hospital	1
Brigham Young University	1
Brookhaven National Laboratory	1
California Institute of Technology	3
Case Western Reserve University	2
Catholic University	1
Children's National Medical Center	1
Cincinnati Children's Hospital Medical Center	1
Colorado School of Mines	2
Columbia University	1
Cornell University	3
Dartmouth	2
Drexel University	1
Duke University	3
Duquesne University	1
FAMU	1

Fermi National Accelerator Laboratory	1
Florida Atlantic University	1
Florida State University	4
Fox Chase Cancer Center	1
Georgetown University	1
Georgia Tech	4
Harvard University	1
HDF Group—a nonprofit that formed out of University of Illinois	1
Illinois Institute of Technology	2
Indiana University	4
Iowa State University	2
Jet Propulsion Laboratory	1
Johns Hopkins University	4
Kansas State University	1
Kent State University	1
Lawrence Berkeley National Laboratory	4
Lehigh University	1
LSU	2
Maine Maritime Academy	1
Mayo Clinic	1
Mayo Institute Biomedical Imaging Resource	1
Medical College of Wisconsin	1
Medical University of South Carolina	1
Memorial Hospital of Rhode Island	1
Memorial Sloan Kettering Cancer Institute	2
Michigan Technological University (Houghton MI)	1
Mississippi State University	1
MIT	6
Montana State University	2
Montana Tech	1
Nathan Klein Institute	1
National High Magnetic Field Laboratory (NHMFL)	4
Naval Postgraduate School	1
NCI	1
North Carolina State University	1

Northeastern University	1
Northwestern University	2
Ohio State University	3
Oklahoma State University	1
Old Dominion University	4
Oregon Health & Science University	1
Oregon Research Institute	2
Pennsylvania State University	7
Princeton University	1
Purdue University	3
Rensselaer Polytechnic Institute	5
Rice University	1
Robert Wood Johnson - Cooper Medical Center	1
RSMAS	1
Rutgers University	2
Sackler Institute Cornell University	1
Saint Louis University	1
Sanford Burnham Medical Research Institute	1
SLAC	1
Southern Methodist University	1
Spallation Neutron Source (DOE)	1
Springfield College	1
Stanford University	2
State University of New York	1
State University of New York Buffalo	1
State University of New York Buffalo—Mechanical and Aerospace Engineering	1
State University of New York Downstate	1
State University of New York Upstate Medical University	1
Syracuse University	1
TERC - non-profit research organization	1
Texas A&M University	5
Thomas Jefferson National Accelerator Facility	1
Tufts	1
Universities Space Research Association—nonprofit research organization	1

University of Alabama Huntsville	1
University of Alabama Tuscaloosa	1
University of Arizona	2
University of California Berkeley	4
University of California Irvine	2
University of California Los Angeles	1
University of California Riverside	1
University of California San Diego	3
University of California San Francisco	3
University of California Santa Barbara	2
University of Central Florida	1
University of Colorado	5
University of Colorado Boulder	5
University of Connecticut	2
University of Delaware	2
University of Florida	9
University of Illinois	3
University of Illinois Chicago	4
University of Illinois Urbana Champaign	1
University of Iowa	2
University of Kansas	1
University of Kentucky	6
University of Louisville	1
University of Maine	1
University of Maryland	1
University of Maryland College Park	2
University of Massachusetts	1
University of Massachusetts Amherst	1
University of Massachusetts Medical School	6
University of Miami	1
University of Michigan	3
University of Minnesota	4
University of Mississippi	1
University of Missouri	1
University of Missouri Rolla	1
University of Missouri-Columbia	1

University of New Hampshire	2
University of North Texas	1
University of Notre Dame	2
University of Oklahoma	1
University of Pennsylvania	3
University of Pittsburgh	1
University of Rhode Island	1
University of Rochester	1
University of Rochester	1
University of South Florida	1
University of Southern California	1
University of Texas	1
University of Texas - Dallas	1
University of Texas APL	1
University of Texas Austin	1
University of Texas Dallas	2
University of Texas Health Science Center Houston	1
University of Texas Health Science Center San Antonio	1
University of Texas MD Anderson Cancer Center	1
University of Texas Southwestern Medical Branch	1
University of Utah	2
University of Virginia	3
University of Wisconsin	1
University of Wisconsin Madison	4
Utah State University	2
Vanderbilt University	5
Virginia Polytechnic Institute	1
Wake Forest University	1
West Virginia University	1
Western Pennsylvania Hospital	1
Worcester Polytechnic Institute (WPI)	1
Wright State University	1
Yale University	1
Yale University School of Medicine	1
Total	294

SOURCE: 2011-2014 Survey, Question 72.

Appendix E

Case Studies

To complement our review of program data, we commissioned case studies of 11 companies that received STTR Phase II awards from the five study agencies. Case studies were an important part of data collection for this study, in conjunction with other sources such as agency data, the survey, meetings with agency staff and other experts, and workshops on selected topics. The impact of STTR funding is complex and often multifaceted, and although these other data sources provide important insights, case studies allow for an understanding of the narrative and history of recipient firms—in essence, providing context for the data collected elsewhere.

Overall, this portfolio sought to capture many of the types of companies that participate in the program. Given the multiple variables at play, the case studies are not presented as any kind of quantitative record. Rather, they provide qualitative evidence about the individual companies selected, and reflect different aspects of the awardee population. The featured companies have verified the case studies presented in this appendix (see Box E-1) and have permitted their use and identification in this report.

ADELPHI TECHNOLOGY, INC.[1]

Adelphi Technology, Inc. is a private company founded in 1984 as sole proprietorship by Melvin Piestrup and incorporated 2 years later in 1986. The company produces a range of high energy neutron sources for industrial and

[1]Primary sources for this case study are the interview with Dr. Charles Gary, August 18, 2015, and a review of the Adelphi web site (http://www.adelphitech.com) and related company documents.

BOX E-1
STTR Company (SBC) and Research Institution (RI) Consultations

Companies

Adelphi Technologies, Inc.; Calabazas Creek Research, Inc.; Creare, Inc.; Ekso Bionics, Inc.; Muons, Inc.; Nanosonic, Inc.; Physical Sciences, Inc.; Stratatech Corporation; Vista Clara, Inc.; Xemed, LLC; Xia, LLC.

A diverse set of case studies were selected for inclusion based on a range of selection criteria: extensive company experience with the program; strong comparative company experience with both SBIR and STTR; company experience with STTR (and SBIR) at a range of agencies; diverse company ownership; a range of company geographical locations--including location in well-known research clusters as well as in less concentrated areas of scientific expertise; and different company commercialization profiles. These profiles range from small research-oriented companies, to larger companies with strong track records as contract research organizations serving specific agency needs, to those focused tightly on the development and commercialization of specific products.

In all cases, appreciation is extended to the executives who took time to participate in interviews and provided further feedback through the review of preliminary drafts.

research applications. Adelphi is headquartered in Redwood City, CA. For its first ten years, the company focused on the research aspects of SBIR/STTR awards, followed by a further ten years in which it was seeking to identify and develop commercial products.

Dr. Charles K. Gary, Vice President for Operations for Adelphi said that his company, in recent years, has completed its evolution from a research-oriented company into a more product-focused company, and at the same time has focused its attention increasingly on the development and then sale of compact neutron generators (CNGs).

CNGs have a number of advantages over isotopes as sources for neutrons: they can be turned on and off, which makes them in practice safer to handle. They eliminate the significant bureaucratic requirements involved in using isotopes, which for instance require a radioactive materials license while CNGs do not. There are no materials handling issues. CNGs can be provided with a relatively small footprint. And isotopes must be replaced much more frequently, for which there are disposal costs. So while the cost of the raw source is much higher for CNGs, the overall life cycle cost is lower.

Reduced bureaucratic costs are especially attractive to academics, according to Dr. Gary, as they do not have the resources easily available to ensure compliance. Hence academic labs have been an important initial market.

The focus on CNGs also open the door to broader use of neutron scattering techniques in research and wider commercialization of neutron-based technologies in both new markets (for Adelphi) such as medicine (as an oncology therapy) and security (as a non-invasive sensing technology).

Adelphi operates an onsite neutron laboratory facility at its headquarters in Redwood City. The laboratory supports Adelphi's own research and development into new generator designs and neutron related applications. The laboratory is also available to customers so they can get first-hand experience with Adelphi neutron sources as they consider incorporating them into their own products.

Adelphi is recognized for its innovative work in the design and development of neutron generators. In 2012, in collaboration with Berkeley's Lawrence Livermore Laboratory, it won an R&D 100 award for their work developing the company's DD100 Series of High Output Neutron Generators. In 2013, in collaboration with the University of Florida, Adelphi won a second R&D 100 award for its DD109X High Flux Fast Neutron Source.[2]

Adelphi maintains research relationships with a broad range of academic, government, and corporate organizations such as the University of California, Berkeley, the University of Florida, Yale University, Indiana University, Rapiscan, Inc., Engility, Inc., and the Savannah River National Laboratory. Adelphi has approximately 10 employees at its headquarters.[3]

Technology: Neutron Sources

Neutron sources are primary used in materials analysis based on neutron scattering. Because neutrons are electrically neutral, they penetrate matter more deeply than electrically charged particles of comparable kinetic energy. They are, therefore, useful sensors of bulk material properties. In scattering experiments, neutrons cause pronounced interference and energy transfer effects. Because they do not interact well with the electron cloud, interference effects stem from neutron-nucleus interactions.

---

[2] "R&D Magazine 2012 R&D 100 Winners," *R&D Magazine,* June 7, 2012, http://www.rdmag.com/articles/2012/06/2012-r-d-100-award-winners; "R&D Magazine 2012 R&D 100 Winners," *R&D Magazine,* July 8, 2013, http://www.rdmag.com/award-winners/2013/07/2013-r-d-100-award-winners.

[3] "Our Teammates," http://www.engilitycorp.com/seaport-e/team-members/.

Until the 1990s, special research facilities were required to generate such neutrons fluxes, either research nuclear reactors or spallation reactors. Researchers applied for beam time to run their experiments at a small group of about 20 research institutions (RIs) globally. The neutron sources developed by Adelphi have much lower capital and operational costs and, although lacking the flux density of these research reactors, are enabling broader use of neutron scattering in research and in industrial applications.[4]

Adelphi neutron sources contain compact linear accelerators that produce neutrons by fusing isotopes of hydrogen together. Deuterium (D), tritium (T), or a mixture of these two isotopes of hydrogen is accelerated into a metal hydride target also containing deuterium, tritium or a mixture. The hydrogen atoms fuse resulting in the formation of helium and a neutron. The energy of the neutrons depends on types of hydrogen isotopes that fused.

The Adelphi technology can produce sufficiently high levels of energetic neutrons for many research and industrial applications. The flux rates of Adelphi's neutron sources are controllable. Also, the flux is monochromatic (if both the accelerated and target isotopes are the same). For example, deuterium atoms fired at tritium targets produce neutrons with uniform kinetic energies of 14.1 MeV.

The principal industrial applications of neutron scattering are in healthcare and security. In healthcare, boron neutron capture therapy (BNCT) is potentially a new therapy for radiation oncologists. In BNCT, boron-10 is delivered to the tumor, either directly via injection or using antibodies. The tumor is irradiated with a neutron beam. The beam does not interact appreciably with tissue. In the tumor, however, boron-10 transforms into boron-11 which is radioactive and kills the tumor cells. Adelphi has already developed proprietary designs for neutron sources in oncology facilities.[5]

Adelphi is also partnering with government and private entities neutron-based scanning systems for application such as border security, airline-cargo inspection, and investigation of unknown packages. Because fast neutrons (> 1 MeV) have deep penetration of most materials—usually over 1 meter—they have significant advantages over x-rays in non-destructive, non-contact scanning.

---

[4]Hammoud, "Introduction to Neutron Scattering," National Institute of Standards and Technology, http://www.ncnr.nist.gov/staff/hammouda/distance_learning/chapter_6.pdf.
[5]"The Basics of Boron Neutron Capture Therapy," http://web.mit.edu/nrl/www/bnct/info/description/description.html.

Business Model

Adelphi Technology has supported operations by performing SBIR research and selling products and services. The company generates approximately $1.5 million annually from the provision of products and services related to the design and development of CNGs, including some SBIR/STTR funding.

Adelphi was initially quite dependent on SBIR funding. However in recent years as more products have reached commercialization, the SBIR/STTR share of total revenue has declined. SBIR/STTR now accounts for about one third of company revenues, according to Dr. Gary, down from well over 50 percent in the early years of the company. He anticipates that this percentage will fall further as markets for CNGs mature, and that Adelphi will receive zero SBIR/STTR funding in 2016.

Adelphi typically sells four to five CNG systems annually primarily to academic customers and government research labs, including significant interest abroad. According to Dr. Gary, units cost approximately $200,000-$300,000 although highly customized models can reach $400,000.

Adelphi is also working closely with potential security and healthcare customers to design Adelphi sources as OEM (Original Equipment Manufacturer) parts in their customers' systems.

Products

Adelphi has designed and developed neutron sources, producing sources with neutron energies ranging up to 14 MeV and output levels of up to 10^{10} neutrons per second. Recently, the company has added neutron detectors to its product line for use in security and healthcare applications.

Deuterium—Deuterium sources

The deuterium—deuterium (DD) reaction produces neutrons sufficiently energetic (2.5 MeV) for non-destructive elemental identification in a wide range of analytic applications. Like the deuterium—tritium sources, these systems consist of an accelerator head, a power supply (2kW) and control rack, and a heat exchanger/chiller. Because deuterium is non-radioactive, Adelphi's DD generators source a continuous supply of deuterium gas from an external tank, resulting in a tube head with almost unlimited lifetime. Other internal components can be easily exchanged by the user as needed due to damage or excessive wear. These generators make excellent fast neutron sources for laboratories and industrial applications that require neutrons with safe operation, small footprint, low cost and small regulatory burden.

Deuterium—Tritium Sources

Deuterium—tritium (DT) sources produce much more energetic neutrons (14.1 MeV) than deuterium—deuterium sources. Thus, DT neutrons penetrate further into objects, for more effective screening and imaging. The DT reaction is 100 times more efficient than the DD reaction, so DT sources have substantially lower operating costs. However, both capital and maintenance costs are higher, and higher energy neutrons require heavier shielding to protect users. Furthermore, because tritium itself is radioactive, the tube head is sealed for user safety. The tritium inside is consumed, and eventually the source must be returned to Adelphi for periodic maintenance, typically after several thousand hours of operation. Also, the customer must register DT sources with the Nuclear Regulatory Commission.

Detectors

Adelphi's detector work has been motivated mostly by the opportunity presented in security applications where the goal is not only to produce neutrons but also to detect their interactions with matter in real time. Detector projects include liquid Argon large volume detectors, a large area scintillation camera, particle imaging, and phoswich detectors for neutron discrimination.

Patents and Other Intellectual Property

Adelphi Technology is the assignee for the U.S. patents listed in Table E-1.

TABLE E-1 Adelphi Technology Patents

Patent Number	Patent	Year
7,177,389	X-ray tomography and laminography	2007
6,992,313	X-ray and neutron imaging	2006
6,765,197	Methods of imaging, focusing and conditioning neutrons	2004
6,674,583	Fabrication of unit lenses for compound refractive lenses	2004
6,545,436	Magnetic containment system for the production of radiation from high energy electrons using solid targets	2003
6,269,145	Compound refractive lens for x-rays	2001
6,201,851	Internal target radiator using a betatron	2001
5,107,508	X-ray laser	1992
5,077,774	X-ray lithography source	1991
4,951,304	Focused X-ray source	1990

SOURCE: U.S. Patent and Trademark Office.

Adelphi Technologies and SBIR/STTR

Between 1984 and 2014, SBIR/STTR funded 91 projects with Adelphi Technology, Inc. amounting to nearly $19.7 million in funding. Of this, DoE accounted for approximately 41 percent, NIH 25 percent, and NSF 17 percent, with the remaining 17 percent from the DoD, NASA, the Department of Homeland Security, and the Department of Transportation. Dr. Gary observed that typically 30 percent of SBIR funding and 40 percent of STTR funding is used for subcontracts.

Adelphi has extensive experience with the DoE SBIR/STTR program. Dr. Gary observed that DoE SBIR/STTR topics were in some cases clearly derived from the research-oriented interests of topic managers, while in others there was a commercial interest as well. Adelphi had initially won a series of more science-oriented awards but as a result of increasing internal focus on commercialization was now more selective in the topics to which it applied. However, some recent awards on neutron optics were in topics that showed limited commercial potential given market realities for that technology.

Dr. Gary was concerned that some topics were simply not funded at all. He believed that DoE should be careful to ensure that topics were excluded from the solicitation if there was no track record of funding. He also suggested that DoE consider funding broader topics. Currently, too many topics are tightly defined technically, which meant that potentially valuable ideas were not considered.

Dr.Gary said that the topic development process at DoE was quite opaque, and he suspected that for a number of topics the process was largely driven by research scientists within DoE. While this resulted in interesting science, he believed that it lacked alignment with commercial opportunities: not all good science is commercially viable.

DoE currently provides one solicitation annually for each broad area of interest; Dr. Gary said that agencies providing more than one solicitation—such as DoD and NIH—were better attuned to the speed of technical development, and that DoE should consider adding at least one additional deadline for solicitations annually.

More generally, Dr. Gary said that connections with DoE staff were very limited. Project liaisons appeared to have other more pressing responsibilities, and in most cases there was almost no contact between the DoE staff and the PI or company representatives beyond the resolution of contracting issues.

In particular, DoE staff were of little help in finding potential markets for the technology within DoE. This contrasts for example with Homeland Security, which clearly considers itself a potential customer for SBIR/STTR products and hence pays quite close attention to progress on the award. Overall, Dr. Gary said that it was very rare to find a DoE program manager who was interested in the funded project; in most cases they simply sought to ensure that no fraud was being perpetrated and that the science was good.

So far as the review process was concerned, Dr. Gary felt that insufficient information was being provided to applicants—in particular, too many applications were graded as excellent but not funded. It would be helpful to have a more granular review that effectively identified weaknesses when projects were not selected.

Dr. Gary was also a strong proponent of better review feedback more generally. He noted that NIH provides an online resource (ERA Commons) where applicants can find all of their applications and all reviews. In contrast, DoE applicants must apply to have a review sent to them, and the window for this application in limited. This substantially reduced the value of the process for the company and imposed unnecessary burdens.

Finally, Dr. Gary wanted to underscore his appreciation for the DoE payments system, which he believed was the best of all the SBIR/STTR agencies. Funding was available immediately and could be pulled in any amount at any time against work and need. This was extremely helpful for a small business, and contrasted very favorably with other agencies that used a milestone-based system.

STTR

Dr. Gary noted that Adelphi typically works with research institutions that are seeking ways to bring their technology to market. In some cases, Adelphi has identified opportunities. In others—for example a current STTR project—the driver is the university where the researcher is the PI. The work in this case is in a fairly esoteric field with minimal commercial potential, but the project has been highly successful technically.

Dr. Gary said that he was a strong supporter of the STTR program, and believed that companies were best placed to determine whether a project should be SBIR or STTR, based on the needs of the project. He observed that a separate solicitation for STTR was likely to generate poor quality partnerships put together primarily to find funding, and that SBIR/STTR should provide a single opportunity for funding.

So far as funding amounts were concerned, Adelphi would certainly consider applying for less funding if there was some benefit for doing so—for example, a higher likelihood of success. As this was not the case for most agencies. The company instead designed the project to meet the funding available.

CALABAZAS CREEK RESEARCH, INC.[6]

Calabazas Creek Research (CCR) is a private company founded in 1994 by Dr. R. Lawrence Ives, who remains President. The company specializes in the design and development of high power electron beam devices, including electron guns and RF sources. In addition to product and service offerings, CCR also licenses software tools for the design of electron beam devices and waveguide components. These software packages simulate particle trajectories, electromagnetic fields, RF fields, thermal performance and RF radiation.

Dr. Ives founded CCR after previously working at for a large defense contractor. While an employee, he reviewed SBIR proposals, and, after starting his company, immediately sought SBIR funding, winning two DoE projects. In both cases, Phase II's were subsequently awarded and provided a foundation for the company in both financial and technical terms—the technology developed for one of the awards is still the most advanced in the world, according to Dr. Ives. The projects also provided a commercial return, with about six sales of devices for testing high-powered gyrotrons, at approximately $120,000 each,

CCR is primarily a research and development firm, developing high power electron beam devices and components for clients working in communications, defense, and particle physics research. CCR employees prototype designs in a laboratory leased from Communications & Power Industries, a $350 million manufacturer of components for the defense and telecommunications sectors.[7]

CCR is a virtual company. Aside from the lab space noted above, it rents or owns no office space. Two employees work in the laboratory and the remaining staff, located across the country, work from home offices. Dr. Ives said that the company's very low cost structure substantially reduces its overhead rate (to slightly more than 20 percent), which allows it to pay wages that are considerably higher than the industry standard. The company offers no paid leave and relies on what Dr. Ives believes to be a much more comfortable and productive environment for its staff.

In addition to providing innovative designs for components in medical and defense systems, CCR provides technology to high energy physics research scientists. For example, CCR partnered with the SLAC National Accelerator Laboratory to improve the performance of cavity resonators in linear

---

[6]Primary sources for this case study are an interview with Dr. Ives on August 21, 2015, and a review of the Calabazas Creek Research web site (http://www.calcreek.com) and related company documents.

[7]Bill Silverfarb, "It is rocket science," *The Daily Journal*, August 15, 2011, http://archives.smdailyjournal.com/article_preview.php?id=165168.

accelerators. Stronger electric fields within the resonators allow shorter accelerators, potentially saving millions of dollars in construction costs.[8]

CCR has received substantial recognition for its work. In 2011 the company received an R&D 100 Award for developing Controlled Porosity Reservoir Cathodes that significantly improve cathode performance and lifespan. CCR leadership has also been deeply involved in strengthening the SBIR program. In 2012, Lawrence Ives received the Champion of Small Business Innovation award for his part in 2011's campaign for the long-term reauthorization of SBIR program funding from Congress.[9]

Because CCR produces world leading technology, its products are in demand outside the United States as well. CCR products can be found in Germany, England, India, Japan, Korea, and China. The company is also developing products to meet DoE's obligations for the ITER project in France.

Technology and Products

Electron Beam Devices

Although semiconductors have displaced vacuum tubes in many logic and communications applications, there remain important niche applications in television transmitters, satellite communications, material processing, defense, and particle accelerators. Calabazas Creek Research designs and develops a broad range of high power, short wavelength devices and components for these applications.

The principle devices produced by CCR include traveling-wave tubes, klystrons, gyrotrons and keystrokes. They operate by modulating a beam of electrons using a mixture of electromagnetic fields and resonance phenomena to generate high power, high frequency RF waves. Although related, these technologies vary in their characteristics and applications.

Much of CCR's work is in the development of klystron and gyrotron technologies. In a klystron, cavity resonators modulate a high energy electron beam with an input signal and convert the resulting modulated beam into an output signal. High performance klystrons operate at power levels to 10s of MW and frequencies up to approximately 100 GHz.[10] CCR has designed RF

---

[8]"SLAC Partners with Small Businesses to Put Technology to Good Use: DOE-funded Program Benefits Companies, the Lab and Society," July 29, 2014, https://www6.slac.stanford.edu/news/2014-07-29-slac-partners-small-businesses-put-technology-good-use.aspx.

[9]"SBTC Honors "Champions of Small Business Innovation,'" February 7, 2012, http://www.nsba.biz/content/printer.4422.shtml. "

[10]"How do klystrons work?" Berkeley Lawrence Livermore Laboratory, http://www2.lbl.gov/MicroWorlds/ALSTool/ALS_Components/RFSystem/.

sources producing RF power from a few milliwatts to 200 MW and at frequencies from a few hundred MHz to 1 THz.

Gyrotrons also feature a cavity resonator. The resonator operates in combination with strong magnetic fields to transfer electron beam energy into RF radiation. This radiation can be formed into a beam and emitted at right angles to the direction of the original electron beam. High performance gyrotrons operate in the 1-2 MW CW range and up to 250 GHz.[11]

As in other electron beam devices, the power of a gyrotron is determined by the energy of the electron beam. Consequently, CCR personnel are skilled in designing different components in these devices (such as electron guns, circuits, collectors, RF windows, *etc.*). Indeed, one of CCR's most successful innovations—the sintered wire cathode, which CCR licensed to Ceradyne—is a sub-component in an electron gun.

Corrosion Mitigation

CCR is now actively working on using atomic layer deposition (ALD) to dramatically improve the corrosion resistance of copper cooling channels (the company has long experience in designing cooling circuits). A current Navy STTR program is focused on this effort, and Dr. Ives believes that this may provide a breakthrough technology with many applications.

This STTR is in partnership with North Carolina State University, and Dr. Ives noted that these kinds of arrangements allow a small company such as CCR to enter entirely new technology areas by tapping into university expertise and equipment. ALD requires equipment that CCR does not have and could not afford, even with a Phase II STTR award, but that is readily available at NC State.

Design Services

CCR provides design and development services for many electron beam devices. Additionally, it also licenses simulation and computational tools that CCR has developed to design such devices more effectively.

Design and Development

CCR offers a range of services related to the design of electron beam devices. Broadly, they are: 1) hardware design, 2) software development, 3)

---

[11]"What is a gyrotron?" Bridge 12, http://www.bridge12.com/learn/gyrotron; E. Borie, "Review of Gyrotron Research," Institut für Technische Physik, August 1991, http://bibliothek.fzk.de/zb/kfk-berichte/KFK4898.pdf.

thermomechanical analysis, 4) electromagnetic analysis, and 5) CAD and other design services. Testing and support services are provided by Communications & Power Industries (CPI)[12] in Palo Alto, CA.

Software

CCR markets intuitive, user-friendly software) for a broad range of electromagnetic and particle simulations to the microwave research community.

Patents and Other Intellectual Property

CCR has historically used patents to protect its intellectual property (IP). (see the list of CCR assigned patents in Table E-2). However, Dr. Ives is concerned that the rising costs of patents, particularly maintenance fees, means that CCR will have to become much more selective about which technologies it seeks to patent.

Dr. Ives was also a strong supporter of the recent DoE initiative to permit companies to spend up to $10,000 per Phase II award for patenting costs. He noted that recent proposed changes in Congress impacting the patenting process would have a highly negative effect on small innovative companies like CCR.

Business Model

CCR is not reliant on SBIR/STTR for revenues. Currently, SBIR/STTR provides about 50 percent of annual revenues, according to Dr. Ives. Its customers have included the U.S. Department of Defense, Department of Energy, the National Aeronautics and Space Administration, Raytheon Company, Titan Pulse Sciences, Inc., NexRay, Inc., KLA-Tencor, Inc., Forschungszentrum Karlsruhe (FZK) (Germany), Communications & Power Industries, LLC., TMD Technology, Inc. (United Kingdom), Japan Atomic Energy Association (JAEA), Stanford Linear Accelerator Center, Naval Research Laboratory, Q-Dot, Inc., ARINC, Inc. Heatwave Laboratories, Inc., Surebeam Corporation, Macrometalics, E-Beam, Inc., Omega-P, Inc., MDS Company, Altair, Inc., H.V. Systems (India), and Samsung (Korea). CCR is also working as a subcontractor to provide an electron gun for a major classified defense program.

---

[12]Bill Silverfarb, ""It is rocket science."

TABLE E-2 CCR Patents

Patent Number	Patent	Year
9,013,104	Periodic permanent magnet focused klystron	2015
8,963,424	Coupler for coupling gyrotron whispering gallery mode RF into HE11 waveguide	2015
8,686,910	Low reflectance radio frequency load	2014
8,664,853	Sintered wire cesium dispenser photocathode	2014
8,547,006	Electron gun for a multiple beam klystron with magnetic compression of the electron beams	2013
7,545,089	Sintered wire cathode	2009
7,313,226	Sintered wire anode	2007
7,102,459	Power combiner	2006
6,987,360	Backward wave coupler for sub-millimeter waves in a traveling wave tube	2006
6,919,776	Traveling wave device for combining or splitting symmetric and asymmetric waves	2005
6,847,168	Electron gun for a multiple beam klystron using magnetic focusing with a magnetic field corrector	2005
6,768,265	Electron gun for multiple beam klystron using magnetic focusing	2004
6,411,263	Multi-mode horn	2002
5,949,298	High power water load for microwave and millimeter-wave radio frequency sources	1999
5,780,970	Multi-stage depressed collector for small orbit gyrotrons	1998

SOURCE: U.S. Patent and Trademark Office.

CCR is also successful in licensing intellectual property developed through SBIR funding. In 2010, Ceradyne acquired the intellectual property rights for "sintered wire" technology that enables the production of a tungsten, reservoir, dispenser cathode with applications in electronic counter measures (ECM), telecommunications, medical devices, defense, and scientific research. The licensed technology improved the cathode current density by a factor of ten and extended cathode lifespan by a factor of two to four times (U.S. Patent #: 7,545,089).[13]

[13]"Ceradyne, Inc.'s Semicon Associates Division Acquires New Ceramic Impregnated Dispenser Cathode Technology," July 26, 2010, http://www.ceradyne.com/news/newsreleasedetails.aspx?id=192.

CCR also generates income by providing design services to the microwave R&D community. Technical services have been provided to numerous organizations, including Karlsruhe Institute of Technology (Germany), Communications & Power Industries, LLC, (USA) Northrop Grumman Corp. (USA), Samsung (Korea), Japanese Atomic Energy Agency (Japan), and SLAC National Accelerator Laboratory (USA).

Collaborations

CCR is strongly oriented toward collaboration, particularly with academic research partners. It maintains research relationships with various academic laboratories, such as the Massachusetts Institute of Technology, North Carolina State University, University of Maryland, and Rensselaer Polytechnic Institute. CCR also works with several industrial organizations, including Ron Witherspoon, Inc. and HeatWave Labs, Inc. Its list of recent collaborators includes:

- University of California—Berkeley
- Rensselaer Polytechnic Institute
- North Carolin State University
- University of Maryland
- University of Wisconsin
- Old Dominion University
- SLAC National Accelerator Laboratory
- Fermilab
- Sandia National Laboratory
- General Atomics
- Los Alamos National Laboratory
- Communications & Power Industries, LLC

SBIR/STTR

Between 1995 and 2014, SBIR funded 119 projects with Calabazas Creek Research, amounting to nearly $31.4 million. Of this, DoE , provided about 75 percent, DoD provided 23 percent, with the balance from NASA and NSF.

STTR

CCR sees STTR as an enormously helpful program and finds that, in some cases, it is a better vehicle for company initiatives than SBIR (in which the company also participates extensively).

Dr. Ives noted that STTR provides an appropriate structure for partnering with research institutions and also offers access to the creativity and enthusiasm of graduate students. A recent STTR with North Carolina State University led to student-developed designs being incorporated into CCR products.

CCR had differing experiences with universities. Some, such as NC State, offered realistic licensing terms and welcomed collaboration with small companies. Others did not appear to understand the limited resources of small businesses and required unrealistic up front licensing fees and royalties. Similarly, there are often complexities in dealing with university technology transfer offices that limit commercialization.

Partnering with research institutions results in other challenges. In particular, universities and students want to publish their research. It was therefore, in Dr. Ives' view, important to understand this need and provide opportunities to publish without compromising company intellectual property. Dr. Ives believes this can be accomplished, as the record of publications related to CCR-university collaborations shows.

Dr. Ives said that when he sees interesting topics in a solicitation that are outside the company's range of expertise, he seeks possible collaborators through his extensive network of technical experts and is often able to identify appropriate collaborators.

Recommendations for SBIR/STTR

Dr. Ives said that none of CCR's major accomplishments would have been possible without SBIR and STTR. He then offered a number of comments and recommendation related to SBIR/STTR, and in particular the DoE SBIR/STTR program, from which CCR receives most of its SBIR/STTR funding.

Topic development. Dr. Ives noted that the wording of topics in some cases did not change from year to year, which in his view suggested that the agency was not interested in these areas.

Unfunded topics. Some agencies appears to publish topics in areas that are unlikely to be funded. These are often topics that appear year after year with no awards being made. This is a waste of time for companies that apply. Topics that are systematically not funded should be eliminated.

Phase III. Most agencies do not have a Phase III policy in place that supports commercialization of technology developed in the SBIR/STTR program. Recent experience with a national laboratory suggests that operations within agencies are not following the Phase III directives in the current SBIR law. Phase III is currently not seen as a responsibility of the SBIR/STTR program office, and it do not appear that it is the responsibility of any other office within the agencies. The exception is the U.S. Navy, which established a Phase III policy and insures it is followed by its operational offices.

More recent focus on commercialization. Dr. Ives said that historically, some agencies appeared to have little interest in commercialization, and that most topics were focused more on addressing technology needs rather than development of commercial products. CCR previously applied for many such topics, and received awards, but realized that it was difficult to build a sustainable business on 6-7 percent profit margins. The company has become much more selective about which SBIR/STTR awards it applies for, with a greater emphasis on commercialization potential.

SBA commercialization benchmarks. Dr. Ives supports the new SBA commercialization benchmarks for awardees with a minimum number of awards. He believes that this will encourage firms to take a more commercial view of their activities.

Letters of intent. Dr. Ives said that the letter of intent (LOI) process provided a good opportunity for companies to explore possible applications without committing substantial resources.

CREARE, INC.[14]

Creare LLC is a private company founded in 1961 by Robert Dean, Dr. Dean was an Assistant Professor of Mechanical Engineering at MIT in the Gas Turbine Laboratory, the Head of Advanced Engineering at Ingersoll-Rand Company from and an Associate Professor and later Professor of Engineering at Thayer School of Engineering at Dartmouth, prior to starting Creare. Dr. Dean is now Professor of Engineering, emeritus. The company is an engineering research and development company, which both acts as an engineering consultancy and commercializes proprietary technologies through licensing or through the creation of independent product companies. Creare is headquartered in Hanover, New Hampshire, and has approximately 150 employees.

Creare is a partnership. It has seven principal engineers who own and operate the company. According to Dr. Rozzi, "for someone who wants to get their technology implemented and see their ideas manifested in the world, it's the ideal place to work—an engineering Disney land."

The company originally provided expertise in fluid dynamics, serving the turbine machinery and nuclear industries during the 1960s and 1970s. In the 1980s, Creare branched out into the energy, aerospace, cryogenics, and materials

---

[14]Primary sources for this case study are the interview with Jay Rozzi, Principal Engineer, Dr. Rozzi's presentation at the National Academies workshop on STTR, May 2015, and a review of the Creare, Inc. website (http://www.creare.com) and related company documents.

processing industries. The 1990s brought growth in software, controls, and biomedical applications. Typical deliverables from an engagement with Creare include analysis with results, experimental data, engineering models, design recommendations, software, numerical solutions, prototypes, and hardware designs.

Although Creare's founding precedes the creation of the SBIR/STTR program, it has proven to be one of the most adept participants in the program. Well positioned by virtue of its capabilities, Creare was able to navigate the early uncertainties in the program because of strong personal ties between then president Jim Block and Warren Rudman, a New Hampshire senator and a key supporter of the original SBIR legislation. Since 1985, Creare has received over 950 awards, $50 million in SBIR Phase I, $197 million in SBIR Phase II, $3.3 million in STTR Phase I, and $10.2 million in STTR Phase II.[15]

Creare's offices and laboratory facilities cover over 60,000 sq. ft. and are located in Hanover, New Hampshire. The office space includes general seating for engineering, technical, and administrative staff, computer facilities, a dedicated technical library, conference rooms and various community spaces. Over half the facility is dedicated to laboratory space, experimental project rigs, machine shops, and specialized fabrication and test apparatus. These extensive facilities and in-house capabilities have been developed and refined over Creare's 50+ year history to serve its broad range of clients. Creare's capabilities enable projects that span development activities in mechanical systems and prototypes, electronics, advanced manufacturing, chemical engineering, nuclear engineering, bioengineering, space-qualified systems, materials development, acoustics, cryogenics, etc. Creare's laboratories are supplied with standardized buses for electric power and pressurized air that enable a broad range of general experimental work. Extensive clean room facilities enable fabrication, assembly, and testing of space-qualified hardware. Its in-house fabrication capabilities are supported by an extensive machine shop and a fully equipped electronics laboratory. To support clients that require qualified and documented hardware, Creare also maintains a quality assurance program and state-of-the-art inspection facilities. Creare's labs are staffed with approximately 40 highly skilled electrical and mechanical technicians, machinists and support staff who typically support approximately 100 concurrent experimental projects in its laboratories.

Creare also maintains research relationships with a broad range of university, government, and corporate R&D organizations. As an example, the list of industry partners working with Creare in the area of advanced

---

[15]"CREARE LLC" https://www.sbir.gov/sbirsearch/detail/263879; National Research Council, *An Assessment of the Small Business Innovation Research Program*, Washington, DC: The National Academies Press, 2008, p. 268.

manufacturing is both long and notable. Creare has strong relationships with machine equipment companies like KMT, MAG IAS, Fives, Harris Aerostructures, Saint-Gobain, Guhring, Iscar, AMETEK/Precitech, among many others. At the same time, it also works with these numerous prime contractors including LMACo, NGC, BHT, ATK, P&W[16] as well as Tier 1 suppliers.

Engineering Services

Creare provides engineering services to a diverse, international customer base, including both government and industrial clients, in a broad range of industries. At present, disciplinary foci include biomedical and human systems, cryogenics, fluid and thermal systems, sensors and controls, advanced manufacturing, and power systems. The following provides a sense of the disciplinary breadth of Creare's engineering work.

Cryogenics

Creare is well known in the areas of miniature high-speed turbomachinery and gas film bearings for cryogenic applications. These specialties are supported by the company's overall expertise in heat and mass transfer, thermal system design and analysis, and the fluid dynamics of multiphase and multi-component flow systems.

Cryogenics projects have included the development of probes for cryosurgical treatment of cancer, superconducting electrical buses for the space station, shipboard liquefaction of helium to cool advanced propulsion systems, and cryogenic cooling systems and packaging for superconducting electronics. Creare also designed, built, and delivered to NASA the cryocooler that fixed the malfunctioning infrared imaging system on the Hubble space telescope. This cryocooler was installed in 2002 and is directly responsible for the over 10-year revival of the NICMOS camera on the Hubble.

Fluid and Thermal Systems

The original disciplinary focus of Creare was fluid dynamics applied to turbines. Long experience in his area provides expertise suitable to any situation, including stationary or rotating machinery, coupled fluid flow, heat, and mass transfer; and chemically reacting flows.

---

[16]Jay Rozzi, "Cryogenic Machining," p. 6.

Projects in this area include maintaining uniform temperatures during integrated circuit operation, evaluating the flow fields at the joints in the Space Shuttle solid rocket motors after the Challenger disaster, developing gas lifts for transporting solids mined in the deep oceans, among many, many others.

Sensors and Controls

Creare projects have included a wireless activity monitor for evaluating movement by patients with certain medical conditions, active noise reduction for communications headsets, and next generation catapult slot width measuring systems for U.S. Navy aircraft carriers.

Advanced Manufacturing

Creare develops advanced materials processing and component fabrication techniques, both as end products for clients and as means to build components for other projects. The main focus is to augment current processes to increase overall affordability and product quality. This work again blends strengths in fluid flow and heat transfer, control systems, hardware, and fabrication. Creare's Advanced Manufacturing Center (AMC) facilities at Creare consist of machine tools, lasers, tool wear measurement systems, tooling, and other associated hardware.

Creare's focus is not only on the development of innovative solutions, but their implementation in a real-world manufacturing environment. In doing so, Creare provides innovative, yet practical solutions for hat enable sustainable quality improvements and substantial cost savings. These key partnerships enable Creare to develop innovative, implementable, advanced manufacturing solutions for U.S. industry. They have designed programs for laser-assisted consolidation of F-35 thermosetting composites (Air Force Phase II SBIR) and laser-based curing of thermoplastics (Army Phase II SBIR). Currently, Creare is working on a large-scale program with the Air Force to transition laser-assisted consolidation to F-35 Wing Skin production. In addition, they have worked with Lockheed, the F-35 program and other key partners to transition Cryogenic Machining for the affordable machining of titanium components for the JSF.

Power Systems

Creare works across the full scale of power systems and related technologies, from detailed design and prototyping of individual components to overall system analyses with thermodynamic analysis of alternative system configurations. This disciplinary area merges corporate competencies in fluid flow, heat transfer, combustion, cryogenics, machine design, and power electronics.

Examples include design and testing of gas turbines based on a recuperated Rankine cycle, design of evaporators and condensers for thermal-to-electric conversion cells, and development of heat exchanger technology for a pressurized-air energy storage system.

Biomedical and Human Systems

Building on core capabilities in precision fabrication, software development, signal and image processing, sensor design, control systems, and thermal/fluid technology, Creare has undertaken various multidisciplinary projects for biomedical clients. Creare frequently works with clinicians at nearby Dartmouth-Hitchcock Medical Center, a 400-bed teaching and research hospital, and at other institutions such as Harvard Medical School, Memorial Sloan-Kettering Cancer Center, and Duke University.

Creare has developed various biomedical technologies including innovative signal processing algorithms and software for cardiac electrophysiology, cryogenic probes for the surgical treatment of cancer, aerosol technologies for mass vaccinations, and robotic control software for performing telesurgery.

As described above, Creare uses its capacity to integrate core capabilities across multiple disciplines. Two technologies described below illustrate Creare's ability to combine capabilities in cryogenics, heat flow, and fluid dynamics.

Cryogenic Cooling of Hubble Infrared Imaging Device

Creare began developing technical capabilities related to cryogenic coolers in the early 1980s, based on one of the company's first SBIR projects. Over 20 years, Creare received more than dozen additional SBIR/STTR projects to develop the technology further. Over the same period, the U.S. government and other clients purchased additional engineering services from Creare that totaled 10 times the magnitude of the initial SBIR funding in this area.

The failure of the cooling system for the infrared imaging device on the Hubble telescope provided an opportunity to demonstrate practical application of this body of technical knowledge. According to NASA, "The Hubble team developed the NICMOS Cryocooler–a state-of-the-art, mechanical, cryogenic cooler that has returned NICMOS to active duty. Using nonexpendable neon gas as a coolant, this closed system delivers high cooling capacity, extremely low vibration and high reliability. It employs a miniature cryogenic circulator to remove heat from NICMOS and transport it to the Cryocooler. The system uses a tiny turbine turning at up to 400,000 rpm (over 100 times the maximum speed of a typical car engine). The NICMOS Cryocooler is virtually vibration-free–which is very important for Hubble. Vibrations could affect image quality in much the same way that a shaky camera produces blurred pictures."[17]

Cryogenically Cooled Machine Tools

Creare has a long history of developing systems for advanced manufacturing. For example, one of its early spin-out companies, Creonics, manufactured controllers for high performance computer numerical control (CNC) machine tools. Linking to its expertise in heat management and cryogenics, Creare developed an integrated system that enabled the effective, indirect cooling of cutting tools with very small flow rates of liquid nitrogen. Implemented in partnership with MAG-ISA Gbmh, this technology enables higher machining speeds (50 percent reduction in cycle time) with equal or improved tool life. For the Air Force F-35 program, Creare estimated potential savings of $300 million from adoption of this technology.[18]

Patents and Other Intellectual Property

Creare is the assignee for 36 patents over the period 1976 to 2015 (see Table E-3).

Business Model

Creare has received extensive support from SBIR/STTR funding. It also generates considerable revenue from engineering service contracts, licensing, and to a lesser extent spin-outs. According to Dr. Rozzi, SBIR/STTR (i.e., non-Phase III work) now accounts for about one-harlf of Creare revenues. Nearly 40 percent of Creare's total revenues come from Phase III commercialization activities related to past SBIR/STTR programs.

---

[17]National Aeronautics and Space Administration, "Small Business/SBIR: NICMOS Cryocooler—Reactivating a Hubble Instrument," *Aerospace Technology Innovation*, vol. 10 no. 4, July/August 2002, http://ipp.nasa.gov/innovation/innovation104/6-smallbiz1.html.

[18]Jay Rozzi, "Cryogenic Machining," http://www.nsrp.org/6-Presentations/Joint/100411_Cryogenic_Machining_Background_and_Application_to_Shipbuilding_Rozzi.pdf, p. 18.

TABLE E-3 Creare Patents

Patent Number	Patent	Year
8,777,529	Mechanism for delivering cryogenic coolant to a rotating tool	2014
8,656,908	Aerosol delivery systems and methods	2014
8,544,462	Systems and methods for aerosol delivery of agents	2013
8,303,220	Device for axial delivery of cryogenic fluids through a machine spindle	2012
8,215,878	Indirect cooling of a rotary cutting tool	2012
8,061,241	Indirect cooling of a cutting tool	2011
8,021,737	Panelized cover system including a corrosion inhibitor	2011
7,954,486	Aerosol delivery systems and methods	2011
7,759,265	Protective cover system including a corrosion inhibitor and method of inhibiting corrosion of a metallic object	2010
7,699,804	Fluid ejection system	2010
7,561,051	Magnet locating apparatus and method of locating a magnet using such apparatus	2009
7,373,943	Self-contained breathing apparatus facepiece pressure control method	2008
7,225,807	Systems and methods for aerosol delivery of agents	2007
7,189,468	Lightweight direct methanol fuel cell	2007
7,183,230	Protective cover system including a corrosion inhibitor	2006
7,100,628	Electromechanically-assisted regulator control assembly	2006
7,053,012	Flexible corrosion-inhibiting cover for a metallic object	2006
6,874,676	Method and structure for welding an air-sensitive metal in air	2005
6,833,334	Flexible corrosion-inhibiting cover for a metallic object	2004
6,794,317	Protective cover system including a corrosion inhibitor	2004
6,444,595	Flexible corrosion-inhibiting cover for a metallic object	2002
6,397,936	Freeze-tolerant condenser for a closed-loop heat-transfer system	2002
6,379,789	Thermally-sprayed composite selective emitter	2002
6,212,568	Ring buffered network bus data management system	2001
6,170,568	Radial flow heat exchanger	2001
6,023,420	Three-phase inverter for small high speed motors	2000
5,938,612	Multilayer ultrasonic transducer array including very thin layer of transducer elements	1999
5,906,580	Ultrasound system and method of administering ultrasound including a plurality of multi-layer transducer elements	1999
5,748,005	Radial displacement sensor for non-contact bearings	1998
5,399,825	Inductor-charged electric discharge machining power supply	1995
5,145,001	High heat flux compact heat exchanger having a permeable heat	1992

	transfer element	
5,033,756	Wide temperature range seal for demountable joints	1991
5,029,638	High heat flux compact heat exchanger having a permeable heat transfer element	1991
4,557,611	Gas thrust bearing	1985
4,357,932	Self pumped solar energy collection system	1982
3,981,540	Rock breaking apparatus	1976

SOURCE: U.S. Patent and Trademark Office.

Spin-Offs

Creare has spun out a total of 10 companies in its history. Examples of such companies include the leading supplier of plasma-based metal cutting systems, Hypertherm, as well as a leading computational fluid dyamics software provider, Fluent, which was acquired by ANSYS in 2006. Although Creare remains a small company, these companies generate over 2000 jobs and half a billion dollars annually.[19] Creare has benefited greatly from these companies' successes. As a general rule, Creare management has provided generous terms for the use of its technology in order to maximize the chances of successful commercialization.[20]

Creare has spun off ten companies during its history, and creating spin-off companies is central to its efforts to commercialization SBIR/STTR developed technologies. Several of the spin-off companies have been purchased by larger firms, e.g. Fluent.

Started in 1983, where Creare used early SBIR funding to develop FLUENT™, a general purpose code for computational fluid dynamics (CFD). Creare says that FLUENT™ became the most widely used CFD code language in the world. The company was spun out in 1988, and was purchased by Ansys in 2006.

The most recent Creare spin-off is Edare, which provides manufacturing and product development services intended to transition innovative technologies into low- and medium-volume production. The

[19] "Cryogenic Machining Technology," http://www.gearsolutions.com/news/detail/7168/cryogenic-machining-technology-from-mag; Jay Rozzi, "Cryogenic Machining Background and Application to Shipbuilding," NSRP All Panel Meeting, October 2011, http://www.nsrp.org/6-Presentations/Joint/100411_Cryogenic_Machining_Background_and_Application_to_Shipbuilding_Rozzi.pdf, p. 4.

[20] National Research Council, *An Assessment of the SBIR Program*, p. 270.

TABLE E-4 A Sample of Creare Spin-offs

Company	Year	Spun Out
Hypertherm	1968	Hypertherm was founded to commercialize plasma cutting technology developed at Creare. Still headquartered in New Hampshire, Hypertherm is now the world's largest manufacturer of plasma cutting tools.
Creonics	1982	Creonics develops and manufactures motion control systems for industrial processes. Acquired by Allen-Bradley in 1990, Creonics is now part of Rockwell International.
Spectra	1984	Spectra is a manufacturer of high speed ink jet print heads and ink deposition systems. Formed around a sophisticated deposition technology developed at Creare, Spectra was acquired by Fujifilm in 2006 and renamed Fujifilm Dimatix.[a]
Fluent	1988	Based on Creare's longstanding expertise in computational fluid dynamics, Fluent began marketing comprehensive computational fluid dynamics software. In 2006 ANSYS Inc. acquired Fluent for $565 million.[b]
Mikros	1991	Based on Creare's advanced electric discharge machining technology, Mikros offers precision micro-machining services.
Verax Biomedical	1999	Verax was founded to commercialize technology to detect bacterial contamination of cells and tissues intended for transfusion and transplantation. They have received seven rounds totaling $28.2 million in venture funding.[c]
Edare	2011	Edare provides manufacturing and product development services intended to transition innovative technologies into low- and medium-volume production.[d]

[a]"Dimatix Acquisition by Fuji Reflects Strong Growth Opportunity For Its Innovative Ink Jet Technology," (June 13, 2006) https://www.fujifilmusa.com/press/news/display_news?newsID=880149.
[b]"ANSYS Signs Definitive Agreement to Acquire Fluent; Broadens Capabilities as a Global Innovator of Simulation Software," (February 16, 2006), http://www.prnewswire.com/news-releases/ansys-signs-definitive-agreement-to-acquire-fluent-broadens-capabilities-as-a-global-innovator-of-simulation-software-55340982.html.
[c]"Company Overview," http://veraxbiomedical.com/company/index.asp; "$28.2M in 7 Rounds from 3 Investors," https://www.crunchbase.com/organization/verax-biomedical.
[d]"About Us," http://www.edareinc.com/pages/about.html; "Edare, Inc." http://www.edareinc.com/pages/about.html.

objective appears to be to provide a home for Creare technologies once demand exists for batch production and beyond. Edare will likely focus on niche products: its first commercial product is VacJac™ Tubing, which provides long life vacuum-insulated tubing primarily. This particular technology does not lend itself to the creation of a standalone spin-off single technology company, nor—because of low volumes—is it well suited to a licensing agreement with a large company. Dr. Rozzi said that the Edare model is therefore focused on building a company that at any one time has two to three programs in production, proving low-medium volume manufacturing typically for government clients (although some commercial clients are also anticipated). This low-volume production may be the end of the transition path for some products, but may also be an important way station on the path to larger volume sales or a licensing agreement once the technology has been fully provide and manufacturing processes rolled out. Dr. Rozzi observed that it is a good model for achieving production of 30 to 50 units, which is hard to do in an R&D environment.

Edare will have two new programs in 2016, according to Dr. Rozzi. One will deliver approximately 40 reduced-footprint swaging machine for the Navy, a project for which Creare will be the prime contractor and Edare will build support and sell those systems to the Navy. The second is not to provide tools to LMACo for noncontact metrology for configuration on aircraft, initially the F-35 Strike Fighter. The system will provide for very rapid noncontact inspections of items such as filled and unfilled fasteners which impact the radar cross-section of the aircraft, replacing current manual procedures.

Licensing

Creare has licensed significant amounts of technology. For example, Phillips Screw Company, AeroVectRx Corporation, Envelop, and MAG-ISA Gmbh have all licensed technology from Creare. Creare has licensed technologies developed in its laboratories such as the cryogenically cooled cutting tool technology now sold by Fives LLC, an spinoff of the former MAG IAS Gmbh, which was acquired by Fives. The exact number of technologies that the company has licensed and the income generated by these licenses, however, is unknown.

Creare often uses multiple funding streams to create new technologies that can have multiple applications, according to Dr. Rozzi. One good example is the development of tools for cryogenic machining of very hard metals, focused on titanium, which used multiple funding streams primarily from Air Force and Navy (along with some additional funding from Army).

The objective was to develop the capacity to machine titanium twice as fast as the current standard. Create met that objective using a new approach and filed multiple patents. The technology is now being commercialized with a partner retrofitting production machines and using the technology to provide new machines as well. Edare is still supplying some of the key components.

Dr. Rozzi said that a direct linear path from Phase I to Phase II to a Phase III transition was very rare. Most technologies—especially those supplied to DoD—required more than just a single Phase II prototype. For example, a measurement device of some kind would almost certainly need certification for production, end user input, multiple iterations, and possibly a qualification process.

SBIR/STTR

Between 1985 and 2015, SBIR/STTR funded 959 projects with Creare, Inc. amounting to over $261 million in R&D support. Of the 96 SBIR/STTR projects awarded to Creare in 2013 and 2014, 73 percent (70 projects) were funded DoD, 22 percent by NASA, and 5 percent by DoE. Over the 30 years of SBIR/STTR funding for Creare, STTR awards account for 5 percent of the total by value.

According to Dr. Rozzi, Creare utilizes SBIR and STTR in the same way: Creare only applies for SBIR or STTR awards if the company can see a clear path to transition and/or commercialization. This could mean developing a specialty product—e.g., the cryocooler for Hubble and other space programs, or the turbo pumps developed for the first Mars rovers with NASA SBIR funding, which have now been adapted for other space program at NASA such as the Curiosity Mars rover. While these are specialized technologies, Dr. Rozzi noted that Creare is exploring more commercial applications for these technologies.

Dr. Rozzi said that in the 1980s, SBIR and STTR was primarily a research program. TPOCs would have pet technology projects, which would typically have no clear path to transition would usually not generate commercial returns. Beginning in the 1990s, this began to change as Industry research and development (IRAD) budgets began to shrink at DoD and at the prime contractors. As these budgets began to decline, SBIR/STTR came to be seen as a more viable alternative for the development of new technologies and new systems at DoD. The shift in the SBIR/STTR program was largely completed in the years immediately after 2000.

Creare makes it a high priority to "get the right people in the room as early as possible—as early as P1 proposal development, "according to Dr. Rozzi. Creare tries to develop the entire team as early as possible, bringing together primes, government people, and technologists. This team-oriented approach has led to considerable transition success.

Working with Primes

Creare has done a lot of work with many primes over the years, according to Dr. Rozzi. He noted that he personally knew many of the Lockheed staff working on the F-35, which for all its issues is making wonderful use of SBIR/STTR to develop technologies that are getting into production. Because Lockheed allocates little funding for R&D to support production, they leverage SBIR/STTR for that purpose. The work now coming under way at Edare to address non-contract metrology originated in discussions with Lockheed, who had encouraged Air Force to publish a topic, under which Creare won an award to develop the relevant technology solution

Creare gets involved in SBIR/STTR solicitations in two ways, according to Dr. Rozzi. In one respect the company has a lot of hammers looking for nails: existing technologies that can be applied to new problems to generate new solutions—the noncontact metrology technology was originally developed for a biomedical MRI application, a new kind of laparoscope to be used for the exact measurement of the location of of tumors during surgery.

Alternatively, the solicitation may generate ideas in entirely new areas. For example, Creare recently won a Phase I award from Navy to develop tools for ultra high speed friction stir welding. The traditional approach has been to use big machines operating at low rpms. Creare is now working to develop a much smaller tool (approximately the size of a router) using much higher rpms (a factor of 20-30 increase in rpm). Creare sees a very large market for this tool given the enormous number of stir welds required both by Navy and other ship builders.

STTR

Creare has worked to developed a network of potential academic partners, and is usually aware of the best RI partner might be. In some cases this is an FFRDC, although the latter usually want full payment of their contract up front, and require approval of a CRADA.

Dr. Rozzi noted that ITAR presented particular challenges in relation to STTR. Creare took a very conservative view of ITAR restrictions, and indicated that it could be difficult to ensure that universities understood and accepted the relevant restrictions, particularly when there were a considerable number of foreign students in most high quality engineering departments.

Dr. Rozzi also noted that there had in the past been conflicts over publishing results. RIs, academics, and graduate students all wanted to publish, and that had in some cases led to conflicts. However, he also noted that said there were ways to publish without breaching disclosure limitations.

Creare STTR partnerships tended to be aligned with schools that were well known to Creare engineers. For example, Purdue was one of the top partners for Creare, and it was also the school from which Dr. Rozzi has received his PhD. The company had also worked closely with MIT in the past, but not so extensively in recent years. Similarly, another engineer had developed a close relationship with the University of Minnesota.

In most cases, Creare directs the STTR project. However, a number of universities have now set up TTOs and incubators for emergent SBC's. Faculty are being encouraged to form companies and work through the incubator. In these cases, they often seek companies like Creare to partner on STTR proposals, but Creare is very cautious about becoming involved in partnerships where the driver is the faculty member, according to Dr. Rozzi.

Overall, the bar is simply higher for Creare involvement in an STTR as opposed to an SBIR. Dr. Rozzi said that unless the RI is a great partner—and some are—money going to the RI will not generate results that are nearly as efficient as Creare doing the work. STTR works best when Creare is seeking access to unique RI technologies—for example, previous STTR with Purdue provided access to modeling for composites machining. The fact that the RI is not is not fireable and not easily made accountable under STTR means that Creare has to be very careful. STTR also required an IP agreement, so if one is not in place, and if Creare does not have existing contacts with the contracts staff at the RI, a considerable amount of work is needed before the proposal can even be advanced. So the partnership really has to be worth it, from Creare's point of view.

Despite these challenges, Creare favors STTR. Working with RIs means that Creare is potentially accessing the best and brightest minds in the United States—Dr. Rozzi sees the program as being like a mini-DARPA, seeking ideas that give the war-fighter an advantage, and believes that STTR has an important role in that over the long term. STTR also offers recruiting benefits, by allowing Creare to work with RI staff and graduate students who are potential employees. Dr. Rozzi said that "we get great people" from these projects.

STTR also differs by agency: Creare did a considerable amount of work for NIH in its early years, especially on hardware of various kinds, but Dr. Rozzi observed that NIH was less interested in hard engineering recently.

Recommendations

Dr. Rozzi said that it might be helpful if the agencies endorsed some of the better model contracts for working with RIs. While some were good to work with, others were very difficult on issues related to IP and payments in particular. He said that this particularly applied to FFRDCs, who were institutionally not interested in SBIR/STTR.

Dr. Rozzi also noted that at DoD in particular, STTR topics tended to be long term and higher technical risk, and that he thought they brought particular value to DoD as a result. Too heavy a focus on immediate commercialization would result in missed opportunities, and he recommended that the agency retain the STTR program and use it to focus on these longer term projects.

EKSO BIONICS, INC.[21]

Ekso Bionics, Inc. ("Ekso") is the wholly owned subsidiary of a publicly traded company Ekso Bionics Holdings, Inc. (OTCQB: EKSO) headquartered in Richmond, California. Ekso was founded in 2005 by Nathan Harding, Homayoon Kazerooni, and Russ Angold, all members of the Berkeley Robotics and Human Engineering Laboratory at the University of California

---

[21]Primary sources for this case study are the interview with Dr. Kurt Amundson, R&D Projects Director, August 21 2015, a review of the Ekso web site, and related company documents and SEC filings. See http://www.eksobionics.com and in particular Ekso Bionic Holdings' 10-K for 2014 at http://www.sec.gov/Archives/edgar/data/1549084/000114420415017256/v403902_10k.htm.

(UCB). The company has had a number of names over the years: first Berkeley ExoWorks, then Berkeley Bionics, and then Ekso Bionics and now Ekso.

Dr. Kurt Amundsen joined the company in 2007 when he was completing his Ph.D at Berkeley. He moved with the company as it expanded in 2008 to meet the need for more space, and then again to the current space in 2012 after releasing its first commercial medical product. He is now the director of the Ekso Labs group.

Products and Services

Using technology licensed from the UCB and augmented by their own work, Ekso designs and markets wearable exoskeletons with applications in healthcare, industrial, and military markets. Users strap an exoskeleton over their clothing to augment their strength, endurance and mobility. Patients with neurological injuries such as strokes can rehabilitate and walk again; industrial workers are able to perform heavy duty work for extended periods; and soldiers can carry heavy loads over long distances.

Ekso has garnered extensive and positive media coverage. Following its series A round in December 2010, the company was *WIRED*'s "Most Significant Gadget of 2010", one of *Time*'s "50 Best Innovations of 2010", and was one of *Inc. Magazine*'s "5 Big Ideas for the Next 15 Years". Media interest remains strong with recent stories from 60 Minutes, Forbes, and National Public Radio among others.[22]

At present, Ekso has two principal business areas: medical technology and engineering services with plans to accelerate go-to-market plans for their Industrial division after recent Equipois acquisition. The Ekso GT is used by hospitals and clinics on patients with lower extremity weakness or paralysis. Through the end of 2014, Ekso had placed over 110 devices (with a revenue value of $12.0 million) in service with over 80 customers. It has licensed its

---

[22] "The 10 Most Significant Gadgets of 2010," December 29, 2010, http://www.wired.com/2010/12/top-tech-2010/?pid=928; Alice Park, "The 50 Best Inventions of 2010," November 11, 2010, http://content.time.com/time/specials/packages/article/0,28804,2029497_2030618_2029794,00.html; Issie Lapowsky, "Meet the Makers of the Wearable Robot," October 30, 2012, http://www.inc.com/magazine/201211/issie-lapowsky/big-idea-4-get-millions-out-of-wheelchairs.html; Bruce Upbin, "First Look At A Darpa-Funded Exoskeleton For Super Soldiers," October 29, 2014, http://www.forbes.com/sites/bruceupbin/2014/10/29/first-look-at-a-darpa-funded-exoskeleton-for-super-soldiers/; Steve Henn, "A Suit That Turns A Person Into A Robot (Sort Of)," June 11, 2015, http://www.npr.org/sections/money/2015/06/11/413406156/a-suit-that-turns-a-person-into-a-robot-sort-of?sc=tw; "How the Exoskeleton Helps Veterans Walk Again," June 21, 2015, http://www.cbsnews.com/news/veterans-affairs-secretary-robert-mcdonald-60-minutes-excerpt/.

technologies to Healthcare Products Gmbh for use in prosthetics and to Lockheed for use in its FORTIS™ exoskeleton.

Ekso also provides engineering services, performing research and development work on military and industrial exoskeletons and related technologies paid for by government grants, "by collaboration partners such as Lockheed, or by engineering services customers such as the U.S. military,"[23] while generating intellectual property. Ekso has had grants from the National Science Foundation, the National Institutes of Health, the Defense Advanced Research Projects Agency (DARPA), and the Department of Defense. This division is cash positive and has generated over 150 international patent cases.

Ekso was taken public on January 15, 2014 through a reverse merger with PN Med Group, Inc., a public medical distribution company. PN Med Group Inc. was renamed Ekso Holdings with Ekso Bionics, Inc. as its sole subsidiary. As of December 31, 2013, the company had 65 full-time employees and 4 part-time employees. Management expects to add between 15 and 20 new employees by December 31, 2015.

EKSO GT™

Exoskeleton systems are highly heterogeneous systems, integrating a broad range of advances in material, electronics, control engineering, sensors, and software development. The Ekso system embodies a variety of innovations. For example, by not requiring power to carry the weight of the exoskeleton, Ekso technology has reduced power consumption for able-bodied exoskeletons by a factor of 1,000. Other larger technology trends that are enabling Exoskeleton technology include ongoing improvements in the energy density of lithium-ion batteries and in on-board computational power as well as cloud based storage for big data and accelerated development of wearable technology.

Ekso's primary product is currently the Ekso GT, "a wearable bionic suit" that provides individuals rehabilitating after "spinal cord injuries, stroke and other lower extremity" weakness the ability to stand and walk (using a cane, crutches or a walker) "with a full, weight bearing, reciprocal gait."[24] Supervised by a physical therapist, the patient walks by shifting of the patient's body to activate sensors that initiate steps. Battery-powered motors drive the legs, replacing the patient's deficient neuromuscular function.

For patients with some motor ability intact (for example, after a stroke or an incomplete spinal cord injury), the Ekso GT allows therapists to teach

---

[23]Ekso Bionic Holdings' 10-K for 2014 at http://www.sec.gov/Archives/edgar/data/1549084/000114420415017256/v403902_10k.htm.
[24]Ibid.

proper step patterns and weight shifts that may allow patients to ambulate more independently.

Clinical evidence suggests that allowing individuals with spinal cord injuries to stand and walk may offer improved post injury medical outcomes and reduce costs. Improvements reported (but not clinically proven) include reductions in readmission, pressure sores and urinary tract infections; improvements in bowel function; and reduced incidence of osteoporosis, pneumonia, cardiovascular disease, and psychological disorder.[25]

In 2012, Ekso delivered its first exoskeleton for medical and rehabilitation purposes. By the end of 2013, the company had introduced four major upgrades to the product, including variable assist software that allows patients to contribute their own power from either leg to achieve self-initiated walking. Another important upgrade was real time data capture that gathers device information during rehabilitation sessions. This improves monitoring of both patient progression and asset utilization.

Causes of serious and permanent limitation in "mobility include stroke, spinal cord injury, cerebral palsy and multiple sclerosis."[26] According to the company's 2015 10-K, the potential market for its medical and rehabilitative products is considerable (see Table E-5)

TABLE E-5 Potential Commercial Markets for Ekso Medical Products

	Total Incidence	Annual Incidence	Percent Estimated as Potential Ekso Users
Stroke		795,000	30
Spinal cord injury	300,000	14,000	80
Traumatic brain injury		285,000	30
Cerebral palsy	764,000	10,000	10-30
Multiple sclerosis	400,000	10,000	10-30

SOURCE: Ekso Bionic 2015 10K.

[25]"Kessler Foundation Presents Data on Ekso™ at World Congress of Biomechanics Another Step Highlighting Robotic Exoskeleton Technology in Rehabilitation," Kessler Foundation, May 7, 2015, https://www.kesslerfoundation.org/media/Kessler%20Foundation%20Presents%20Data%20on%20Ekso; "Ekso(tm) clear leader in exoskeleton comparison study," October 27, 2014, http://www.medica.de/cipp/md_medica/custom/pub/content,oid,49060/lang,2/ticket,g_u_e_s_t/~/Ekso_tm_clear_leader_in_exoskeleton_comparison_study.html.

[26]Ekso Bionic Holdings' 10-K for 2014 at http://www.sec.gov/Archives/edgar/data/1549084/000114420415017256/v403902_10k.htm.

Ekso is aggressively pursuing the stroke and spinal cord injury (SCI) rehabilitation segments of this market. The company uses a direct sales force (nine sales people, eight clinicians, five marketers, and three customer service personnel) to reach inpatient and outpatient centers providing stroke and SCI rehabilitation in the United States, Canada, the United Kingdom, Spain, and the German-speaking countries of Europe. Ekso sells to Mexico, Italy, Poland, Turkey, Scandinavia, Ireland, and the UAE via distributer network.

The sales process is complex with multiple stakeholders including among others the clinic CEO/CFO, the medical director, the clinical staff, patients, and the fund-raising director. To build consensus among these stakeholders can take from 3 to 18 months, according to Dr. Amundson.

Because the market of rehabilitative exoskeletons is a new market, the FDA has evolved in how they classify these systems. Ekso initially registered the Ekso product line with the FDA as a Class I 510(k) exempt Powered Exercise Equipment device. In June, 2014, the FDA announced a new product classification for Powered Exoskeleton devices, and in October, 2014, the FDA determined that this new product classification applied to the Ekso GT device. Ekso resubmitted its 510(k) application in December, 2014 and has been collaborating with the FDA to meet the class ll requirements.

Other companies are also producing exoskeletons Cyberdyne, ReWalk Robotics, and Rex Bionics sell ambulatory exoskeletons. Hocoma, AlterG, Aretech and Reha Technology sell treadmill-based gait therapies. All are potential competitors to Ekso in its core rehabilitation segments.

Engineering Services

In addition to further developing it rehabilitation exoskeleton, Ekso is developing systems for able-bodied applications of the technology. In such applications, exoskeleton-enabled individuals become stronger and able to undertake greater effort for longer periods with greater safety. These projects are funded by grants from government, principally the military, and from corporate partners.

Ekso developed an able-bodied exoskeleton, the Human Universal Load Carrier (HULC), in 2008. Designed to enable users to carry up 200 pounds of materiel over long distance and rough terrain, the HULC used Ekso' low power load carriage technology and hip actuation to assist in moving the user's legs during walking.

Further development of able-bodied, powered and nonpowered exoskeletons builds on the HULC technology, and Ekso has developed a nonpowered exoskeleton called Ekso Works. Industrial workers wearing this exoskeleton will perform their tasks with reduced risk of injury from lifting and working with heavy equipment and increased productivity. The company is also undertaking field tests of its exoskeleton in steel production and concrete pouring. Commercialization could begin as early as the fourth quarter 2015.[27]

For the military market, as part of Tactical Assault Light Operator Suit (TALOS) project, Ekso was also recently awarded a pair of contracts by the U.S. Special Operations Command (SOCOM) to design, build, test, and deliver a next generation military exoskeleton prototype. The goal of project is to develop a self-contained, bullet-proof suit that will provide the wearer superhuman strength. Ekso also participated as a sub-contractor to Boston Dynamics, a company owned by Google, in DARPA's Warrior Web program. Ekso has also participated in a competitive test of exoskeletons over an 84 mile obstacle course that was sponsored by the Army Research Laboratory.[28]

Ekso has limited rights to commercialize these able-bodied technologies. In the industrial segment, the company and Lockheed have co-exclusive rights, with Ekso having the right to sublicense technology and Lockheed having the right to sublicense only with Ekso's consent. In the military arena, "Lockheed and the Company have co-exclusive rights to military markets through 2017. So long as certain annual minimum obligations are met, Lockheed will obtain exclusive rights to the government market after 2017."[29]

At present, DARPA, and U.S. Special Operations Command (SOCOM) are the principal customers for contracted engineering services. In 2014 and 2015, SOCOM contracted for $3.1 million in services for its TALOS project, as a sole source follow on to an STTR award.

Patents and Other Intellectual Property

Ekso owns a patent portfolio covering medical exoskeletons, commercial exoskeletons, actuators, and strength enhancing exoskeletons. The portfolio includes six U.S. patents (4 of which are co-owned with the Regents of the University of California), 10 U.S. patents exclusively licensed from the Regents of the University of California, and 6 patent applications currently pending. Thirty-seven applications have been issued as patents internationally in different countries.

---

[27] "Ekso Bionics™ Announces Launch of Ekso™ Labs," March 28, 2014, http://ir.eksobionics.com/press-releases/detail/64/ekso-bionicstm-announces-launch-of-eksotm-labs.

[28] Stew Magnuson, "SOCOM's 'Iron Man' Suit Faces Major Technological Hurdles," January 28, 2015, http://www.nationaldefensemagazine.org/blog/lists/posts/post.aspx?ID=1725.

[29] Ekso Bionic Holdings' 10-K for 2014 at http://www.sec.gov/Archives/edgar/data/1549084/000114420415017256/v403902_10k.htm.

TABLE E-6 Ekso Patents

Patent Number	Patent	Year
9,011,354	Hip and knee actuation systems for lower limb orthotic devices	2015
8,968,222	Wearable material handling system	2015*
8,945,028	Device and method for decreasing energy consumption of a person by use of a lower extremity exoskeleton	2015
8,894,592	Device and method for decreasing oxygen consumption of a person during steady walking by use of a load-carrying exoskeleton	2014*
8,801,641	Exoskeleton and method for controlling a swing leg of the exoskeleton	2014*
8,231,688	Semi-actuated transfemoral prosthetic knee	2012*

*Patents marked with an asterisk are co-owned with the Regents of the University of California.
SOURCE: U.S. Patent and Trademark Office.

The core of Ekso's patent portfolio consists of two license agreements and one amendment with the University of California covering 10 patents exclusively licensed to the company. In consideration for these rights, Ekso paid $5,000 in cash, gave the university 310,400 common shares of Ekso, and currently pays a 1 percent royalty on all sales (excluding products sold or resold to the U.S. government).

Funding

Ekso has relied on a mixture of revenue, debt, equity, and government support to commercialize Ekso GT and develop the other exoskeletons in its product pipeline.

Venture and Angel Investment

Ekso received a considerable amount of early angel funding, according to Dr. Amundson, before turning to the venture community for support in commercializing its exoskeleton technology. A series A round in December, 2010 was followed in June, 2012 with a $9.1 million series B.[30] The company

[30] "UNITED STATES SECURITIES AND EXCHANGE COMMISSION FORM D," (June 22, 2012) http://www.sec.gov/Archives/edgar/data/1552921/000155292112000001/xslFormDX01/primary_doc.xml.

also a received a NIST ATP award that Dr. Amundson said was "critical to the company—we would not be here without it," as a hardware-intensive program like Ekso's requires substantial funding.

Initial Public Offering

Following a year of rapid growth, in Q3 2013 the company was unable to raise a third round of venture capital. The company undertook a reduction in force to reduce its burn rate prior to its IPO. Bridge financing of $5 million in debt enabled the company to become a public company in January 2014 through a reverse merger with PN Medical Group. The IPO raised $22.8 million.[31]

Licensing

Ekso has had success licensing its technologies. Since 2008, the company has received over $1 million in licensing fees. These include upfront licensing fees from Lockheed for military exoskeleton technology and from OttoBock Healthcare Products Gmbh for technologies used in prosthetics and related areas.

Ekso and SBIR/STTR

Between 2005 and 2014, SBIR/STTR funded 13 projects worth $5.03 million with Ekso and its predecessor Berkeley Bionics. Of this, DoD provided about 65 percent, NSF 32 percent and NIH the remaining 4 percent. Forty-six percent of these funds were STTR and the balance SBIR.

Dr. Amundson said that while he very much appreciated the funding available through the SBIR/STTR program, he had been surprised by the lack of contact with program managers. While they had been helpful when contacted, he believed that they were severely constrained by limited administrative funding: NSF had provided $3 million in funding but no NSF program manager had visited the Ekso facility. He believed that especially when making

[31]"UNITED STATES SECURITIES AND EXCHANGE COMMISSION FORM D," (November 27, 2013) http://www.sec.gov/Archives/edgar/data/1552921/000155292113000001/xslFormDX01/primary_doc.xml; "UNITED STATES SECURITIES AND EXCHANGE COMMISSION FORM D," (December 4, 2014) http://www.sec.gov/Archives/edgar/data/1549084/000114420414072378/xslFormDX01/primary_doc.xml.

significant long-term investments, it would be prudent to meet more frequently in person.

Dr. Amundson also noted that Congress appeared to have little understanding of SBIR/STTR . He wondered whether it might be possible for SBA or the agencies to undertake regular regional tours of SBIR/STTR winners for Congressional staff and representative or senators, to demonstrate directly the impact of funding the benefits that flowed from it.

Dr. Amundson said that DCMA auditing practices had caused considerable difficulty. The requirement that small companies show financial sustainability (i.e. profitability or close to it) for two full future years before they could be approved to receive an award caused difficult for many companies. Ekso was still losing money, and although it still had substantial capital available, that would not guarantee financial coverage for two full years. Other small companies would he believed be in even more difficult circumstances.

Dr. Amundson also noted that his experience with NIH had been challenging in part because the agency—or its selection panels—did not seem to accept that engineering was an important part of innovation, and the while the exoskeletons developed by Ekso were no longer novel in the biomedical sense, they still required substantial and expensive engineering before they could reach the market.

STTR

SBIR and STTR are close to identical, except for the need for an academic partner under STTR, according to Dr. Amundson. Which program is utilized is largely determined by the solicitation.

STTR was especially helpful at the earliest stages of company development, Professor Kazerooni was (and still is) on the faculty at UC Berkeley, and STTR was an important bridge from the university to the company. Key employees (beyond the founders) also came out of the lab at the university. So STTR was initially the perfect bridge program.

As the company had matured and developed new sources of funding, STTR has become somewhat less attractive, Dr. Amundson observed. The company now wants to move fast and is less interested in academic research. It does not wish to be so tightly coupled to university needs and interests.

Ekso currently has an STTR award in conjunction with SRI, which is providing to be a more flexible kind of research institution partner. Ekso is now using SRI to do component level technology development that Ekso does not wish to do (or cannot do) in-house. This frees Ekso resources for more focused work on upcoming products. More generally, Dr. Amundson noted, STTR gives the company a way to adapt research done elsewhere, leveraging for example some of the work that DARPA and others have done with SRI.

NSF funding included two $500,000 STTR Phase II awards and a Phase IIB. These provided steady funding over a number of years to support technical development at the company. More recently, Ekso has received STTR awards from SOCOM at DoD, for work on the TALOS program. The TALOS program has funded two STTR awards, one based on a solicitation and a subsequent sole source award.

MUONS INC.[32]

Muons, Inc. (Muons) is a small private technology company based in Batavia, Illinois, with a wholly-owned subsidiary, Muplus, Inc., that is incorporated in Newport News, Virginia. Muons offers a range of products and services, with a primary focus on particle accelerators for high-energy and nuclear physics discovery science, for secondary beams, and for nuclear power. The company currently typically has between one and two dozen employees, and is owned by its founder, chief scientist, and President Rolland Johnson, who has been involved in particle accelerator research and development for over 40 years.

Dr. Johnson said that he started the company to help DoE accomplish its goals through the SBIR program, which was originally created to allow industry to contribute its intellectual and creative energies to national programs in most branches of the government. Having worked in the national labs for many years, he believed that Muons could do things for the labs that needed extra creativity and more funding. Muons hired the most creative people it could find, who were often near national laboratories and who were unable to relocate.

Muons is very different than other SBIR-STTR companies. Dr. Johnson said that most of its work is providing ideas and concepts for national labs, focusing on identifying projects and technologies that will help the labs, but for which there is no available funding, while other companies mostly transfer technology in the other direction. STTR in particular has been used to meet those needs, perhaps acting as a DoE analog to Lockheed's famed Skunk Works as a source of innovative technologies.

Muons has always had close connections to the National Labs. Dr. Johnson spent most of his career at National Labs, initially Lawrence Berkeley National Laboratory (LBNL) and then Fermi National Accelerator Laboratory (Fermilab), where he worked for 17 years before moving to the private sector to install and commission the CAMD synchrotron light source at LSU and then to the Thomas Jefferson National Accelerator Facility (TJNAF) in Newport News where he also served as a detailee at the Department of Energy in Germantown, MD. After retiring, he built a consulting practice and in 2002 founded Muons. The company's first STTR award was in 2003. Since then, Muons has received

---

[32]This case study draws primarily on materials published by Muons on the company's web site, an interview with Rolland Johnson, CEO and Founder, August 27, 2015, and other company materials.

24 Phase II SBIR and STTR awards, and is the largest recipient of STTR awards from DoE.

From its founding in 2002 until 2010, Muons mainly focused on muon collider particle research, and on developing related new technology. It used consulting contracts and SBIR-STTR awards to fund this work. In 2010, the company started exploring Accelerator Driven Subcritical Reactors (ADSR), and this has become a thrust of its commercialization efforts.

Muons workforce varies according to the SBIR-STTR and contracts they are awarded, where fluctuations are mostly accommodated by the number of postdoctoral employees they are able to hire to train in accelerator science who often move on to permanent jobs in national labs. Muons also hires post docs who work within research partner national labs while supported by the company. Muons supports PhD students working on SBIR-STTR grant topics, where three women and one man received their degrees in the last two years. The company is best viewed as primarily a research organization, developing cutting edge technology, although Muons has recent shifted to become more commercially oriented, as has been required by the most recent SBIR-STTR reauthorization legislation. The most significant commercial application is GEM*STAR.

GEM*STAR: Accelerator-driven Subcritical Reactor for Improved Safety, Waste Management, and Plutonium Disposition

Muons has formed and is leading the GEM*STAR Consortium of four companies (Muons, ADNA Corp., Niowave, Inc. and Newport News Shipbuilding), two national laboratories (ORNL and TJNAF), and two universities (Virginia Tech and George Washington University) and has submitted a proposal to DoE Nuclear Engineering for a $50 Million, 5 year funding opportunity titled "Advanced Reactor Industry Competition for Concept Development".

GEM*STAR is a transformative and disruptive technology that has the potential to revitalize the nuclear power industry and lay the groundwork for a path to a viable future for many centuries. It combines proven technologies to provide a new approach to the safety of nuclear reactors, to the management of nuclear waste, and to the disposition of nuclear weapons materials. The primary technologies involved, a molten-salt reactor and a high-power proton accelerator, are not new and have already been proven in the Molten Salt Reactor Experiment at ORNL and at many accelerator facilities around the world. It is designed to be commercially profitable and politically adoptable.

It can burn spent nuclear fuel, natural uranium or thorium, depleted uranium, and surplus weapons material, all without isotopic enrichment or chemical reprocessing. Burning the waste from current reactors can potentially extend their lifetime and turn a huge liability into highly profitable use. Interestingly, with a fleet of accelerator-driven systems like GEM*STAR there is enough uranium out of the ground today to supply the current U.S. electrical power usage for more than 1,000 years. Burning the spent nuclear fuel from the current fleet of nuclear reactors is vastly superior to throwing away its enormous internal energy and just piling it in a hole in the ground for 100,000 years.

Safety: Being subcritical, fission stops when the accelerator is switched off and passive air cooling is sufficient to maintain safe reactor temperature. The system design avoids the major problems associated with all of the historical reactor accidents involving radioactive releases.

Nuclear Waste and Pu Disposition: The accelerator beam generates an enormous neutron flux that induces fission power to generate heat and to transmute fission products and heavy actinides into more tractable waste products. The waste stream from GEM*STAR systems is less of a burden on an ultimate geological store than current reactors, and recycling the waste stream in other GEM*STAR systems could potentially make such a store unnecessary. That same neutron flux burns surplus weapons-grade plutonium more completely than other approaches and satisfies the goals of the year 2000 Plutonium Management and Disposition Agreement between the United States and Russia to each dispose of 34 metric tons of weapons-grade plutonium (enough for 17,000 Hiroshima-sized bombs).

Nuclear Weapons Proliferation is addressed by GEM*STAR operation in that neither isotopic enrichment nor reprocessing is required and by its application to destroy nuclear weapon materials.

The Pilot Plant to be designed will first burn natural uranium as a test and then be upgraded to a mission-capable system for disposing of surplus weapons-grade Pu. The heat generated will be used to drive the Fischer-Tropsch process to provide green diesel fuel to the U.S. military at a profit. This approach mitigates some regulatory issues and avoids the need for initial availability to meet the demands of the electrical grid. This project will carry GEM*STAR through completion of the Conceptual Design Report and the Technical Design Report, including engineering drawings sufficient for the licensing process and to begin pilot plant construction. Experimental studies to improve the design will also be performed.

Muons Technologies

While Muons pivoted in 2010 to focus on ADSRs, they are still developing other technologies including:

- Numerical Simulation Programs and Graphical User Interfaces to them
- RF technology, both normal and superconducting
- Magnetron power sources
- Superconducting magnets for high fields and high radiation environments

Muons' particle physics simulation programs, G4beamline and MuSim, can be used across the particle accelerator industry. G4beamline is free, open source modelling software based on the GEANT4 program developed by a large collaboration headed by CERN and SLAC that accurately simulates the interactions and decays of subatomic particles. According to Muons' website, G4beamline is downloaded ~15 times weekly, and given the small population of potential users, that accounts for a sizeable percentage of global demand. MuSim is a new particle accelerator simulation program that Muons will license that interfaces to GEANT4 and to MCNP, the workhorse of the nuclear physics community.

Muons also develops technologies that use advanced **Radio Frequency (RF) technology**, including the superconducting resonant cavities that accelerate particles by using microwave electromagnetic fields. These cavities are usually powered by klystrons or Inductive Output Transmitters (IOT). Magnetron power sources, based on the same technology as ordinary kitchen microwave ovens, have the potential to be more efficient and less costly than the klystrons or IOTs if they can be made more frequency and phase stable and controllable. Muons has several magnetron projects underway that are based on new ways to stabilize and control magnetrons that can reduce the cost of RF power sources for accelerators by as much as a factor of five and improve efficiency from 50 percent to 90 percent compared to klystron sources. This could make Muons products attractive commercially for a number of applications such as production of radioisotopes for medical diagnostics and therapies.

Superconducting magnets. Muon colliders require a high level of muon beam cooling to work effectively. Muon cooling depends on strong and efficient superconducting magnets, which Muons also develops. These magnets are extremely demanding, as some of them need to create extremely strong magnetic fields in complex shapes with forces that require sophisticated engineering.

Electron Recirculating Linear Accelerators. Muons is working on Electron Recirculating Linear Accelerators (RLA) to make radioisotopes for diverse applications such as those used for diagnostics and therapy in nuclear medicine. Muons is developing new techniques for developing commonly used isotopes as well as isotopes for new medical and industrial applications.

Business Model and Customers

Muons is a small research oriented firm with changing commercial ambitions. Its funding was in large part derived from SBIR-STTR awards, along with some consulting revenues mostly from national labs. However, the company has recently shifted to become more commercially oriented.

Introduction of the new SBIR/STTR commercialization metrics after reauthorization nearly bankrupted Muons, according to Dr. Johnson. In 2011-2 the company was designated as not commercial and hence SBIR/STTR funding dried up, leading to lay-offs.

However, the company has ramped up its commercial activity with contracts from Fermilab to develop plans to upgrade one of their flagship experiments and Toshiba and Niowave to build magnetrons. The company is close to delivering its first commercial magnetron prototype for Niowave, and expects to provide a testable product that delivers a substantial upgrade in power, from a previous tetrode maximum of 60-70KW to more than 120KW. Besides contracts with its usual research partners, Muons has won non-SBIR-STTR contracts with Los Alamos National Lab and Pacific Northwest National Lab. Non-SBIR-STTR contracts have generated almost $2 million in revenues, mostly in the past 5 years, according to Dr. Johnson.

As a result of these efforts, Muons and MuPlus are now seen by the DoE as commercial companies eligible for SBIR and STTR awards, and have won four in the past year. MuSim, mentioned above, is an important non-STTR project, according to Dr. Johnson. Since it interfaces to many simulation tools including MCNP6, it will be extremely useful to develop the Conceptual and Technical Design Reports that are needed for the GEM*STAR project described above. Muons originally developed a similar tool, G4Beamline, which was provided free and is now in use by many companies and labs. Dr. Johnson said that Muons was able to identify over $18 million in effort generated by the program and he believes that MuSim will have an even larger user community of Nuclear Physicists and Engineers who need a better interface for MCNP6. Muons plans to charge for the MuSim program and is spinning out a new business in software support.

Muons Partnerships

Muon partners with multiple third parties on many of its projects. A proposal for a muon beam cooling experiment for example listed 40 individual collaborators and 5 other institutions. The GEM*STAR proposal has seven partner institutions. Muons has partnered with 9 national labs:

- Argonne National Laboratory
- Brookhaven National Laboratory
- Fermi National Accelerator Laboratory
- Thomas Jefferson National Accelerator Facility
- Los Alamos National Laboratory
- Lawrence Berkeley National Laboratory
- National High Field Magnet Laboratory
- Oak Ridge National Laboratory,
- Pacific Northwest National Laboratory

Muons has an especially close partnership with Fermilab, where ideas for muon cooling for colliders, neutrino factories and stopping beams have been developed and TJNAF, where the newest interest is the development of concepts for electron-ion colliders.

Muons has also partnered with eight universities: Cornell University, University of Chicago, Florida State University, Hampton University, Illinois Institute of Technology, North Carolina State University, Northern Illinois University, and Old Dominion University.

STTR

Muons has received 56 DoE Phase I awards, and 24 Phase II awards. 36 of these awards are SBIR, and 44 are STTR. Total funding (2002-2014) is about $26 million.

Dr. Johnson observed that most companies do not want to deal with STTR grants: "We are masochists, since most companies do not want to deal with National Lab bureaucracies and do not want to share their grant money with the lab. However, most Muons staff members are located near the labs where they used to work, and are often embedded in the labs which give them work space. The Cooperative Research and Development Agreements (CRADAs) that are sometimes required for STTR grants with National Labs often include a section detailing how the labs will make available specific lab and office space".

The company first used STTR grants to develop new ideas for a muon collider, addressing the technical problems of cooling beams of muons so they were are dense enough to make such a machine possible. Muons subsequently branched out to related technologies and then some less related areas. The company is now using STTR grants to work on an electron ion collider using polarized electrons and ions at TJNAF. Dr. Johnson believes that this project may have significant commercial potential, although development is still at a very early stage and it takes a considerable time to move from an idea to a product. He noted that this leads to tension inside the DoE SBIR-STTR program, which seems to be seeking commercial outcomes soon after the conclusion of a Phase II award. He noted that a typical time from conception to start of payback in large commercial enterprises is more like nine years.

Dr. Johnson said that DoE STTR grants used to require a CRADA, but they are now structured more flexibly, and require only an IP agreement with the Lab (this is part of the CRADA). The STTR grant also requires approval from the DoE Cognizant Officer who is responsible for lab activities, which can also take considerable time. Currently, most labs that use CRADAs require that separate CRADAs be signed for each of the two award phases, which lengthen delays and adds cost. Each CRADA specifies a time period for work to be completed, and amending this requires a change to the CRADA, as does any other significant change to the statement of work (e.g. a shift to a different part of the lab as provider of a device or service).

Dr. Johnson noted that STTR projects can only work well if there is goodwill between the lab and the company. Because Muons has such long and deep connections with national labs, its staff know most of their counterparts at the labs, so the connection is always positive.

Still, lab administrators in general tend to view STTR awards as small projects. From a $150,000 award, the lab will receive maybe $50,000-60,000, and it costs them almost that much just to do the paperwork, according to Dr. Johnson. So STTR agreements can take a long time to receive signoff from the labs, as they are a low priority for lab administrations.

In some cases, these delays mean that the labs and the company are out of sync, and that the lab will struggle to provide its deliverable on schedule. If a lab fails to deliver on time, the company has to step in to fill the gap, which can cause considerable hardship and economic losses for the company. Namely, the company then has to pay for the work directly yet ends up paying the lab anyway as part of the binding STTR agreement.

DoE program managers are quite flexible, but are constrained by STTR legislation which requires that the Research Partner Institution receive a minimum percentage of the award. Program managers will sometimes allow a switch of RI, but in reality this is not practical: the RI has usually been selected because of its specialized expertise. Dr. Johnson said that program managers should be given the flexibility to switch STTR funding back to the company in special circumstances.

Dr. Johnson said that the DoE STTR-SBIR programs run very smoothly. Recent changes, such as the introduction of letters of intent to allow reviewers to be selected in good time and the well-designed timeline on the agency web site, are welcome improvements.

NANOSONIC, INC.[33]

NanoSonic, Inc. is a small nanotech company based in Blacksburg, VA. Founded as a spinout from Virginia Tech's College of Science and Engineering in 1998 by Dr. Richard Claus, an electrical engineer, it currently has about 35 employees. The company is managed by President Dr. Jennifer Lalli, CTO Dr. Vince Baranauskas, CFO Melissa Campbell, and CEO and Director of Advanced Development Dr. Richard Claus.

Nanosomic was formed to design and manufacture innovative materials, especially new materials that are unavailable in the commercial

[33]The primary sources for this case study are an interview with Dr. Jennifer Lalli, CEO, August 25, 2015, the NanoSonic Inc. web site (http://www.nanosonic.com/), and other materials from NanoSonic.

market. A major company objective is to create these innovative materials through environmentally benign processes and techniques.

Originally, the company focused on the fabrication of thin films and sensors, but soon expanded its activities to include the scale up of coatings and the use of specialized coatings for a range of applications, according to Dr. Lalli. The company hired several chemists to pursue these new areas, and is now concentrating on materials production rather than electronic products

SBIR/STTR awards led to a considerable amount of positive press, Dr. Lalli noted, and this led to more awards and then on to three separate Phase III contracts within three years. The first Phase III award was transformative, as it helped NanoSonic scale up manufacturing production very substantially. As the existing facility in Blacksburg was not suitable, this led to a shift to a new facility about 15 miles from Virginia Tech. The new building was not attached to any other buildings, so provided the added benefit that NanoSonic could perform classified work. More recently, NanoSonic has been seeking to take products to the demonstration stage as early as possibly, and then to move forward to cut costs and scale production rapidly.

NanoSonic's innovative materials have attracted considerable interest especially from DoD prime contractors, who have often heard of NanoSonic through the SBIR/STTR program, according to Dr. Lalli. The company is experienced at putting materials through quality testing, and can provide materials as almost or fully qualified products for bulk of sales to defense primes.

Dr. Lalli said that overall, NanoSonic has had more success selling to primes than to DoD itself. She noted that while SBIR and STTR topics and subtopics supported the development of advanced materials, unless DoD had written a specification for them, there was little likelihood that they would be adopted by the agency: without a new specification, existing materials would continue to be used instead. In that respect, the SBIR/STTR topics were often well ahead of agency procedures.

These difficulties with DoD has led NanoSonic to take a strategic decision to work more closely with the prime contractors, and to de-emphasize efforts to sell directly to DoD, where NanoSonic in the past has had success (on two projects) in using the sole source designation that comes with SBIR/STTR awards.

NanoSonic has made no effort to raise third party funding, even though NanoSonic's metal rubber products had attracted VC interest, in part because the company is able to bootstrap growth through sales and in part because venture funding entailed potential risks.

The company works with all different sizes and types of companies and organizations, and clients include NASA, DoD, and DOT, providing services that cover all phases of product development; R&D, design and development, and manufacturing. R&D services cover polymer and small molecule synthesis, protective coatings, advanced textiles, antennae, RF testing, and sensors.

Technology and IP

NanoSonic's R&D lab is equipped for the design and synthesis of material precursors (compounds that are formed into other compounds through chemical reactions). The lab also forms synthesized precursors into thin (between 1 nm and 1 μm) and thick film materials, using advanced computers for material design, device modeling, and data analysis. The manufacturing lab is mainly dedicated to HybridSil® and HybridShield® production—it produces 4,000 lbs/ day of HybridSil® and HybridShield® nanocomposite formulations.

The company has licensed nine patents from Virginia Tech, covering electrostatic self-assembly processing and use, and is establishing its own intellectual property portfolio in the next step toward commercialization. Currently, NanoSonic has one patent that generally relates to self-formation of a transparent, abrasion-resistant optical coating on solid plastic substrates that protect a solid substrate from wear and/ or provide properties such as magnetism, electrical conductivity, and UV absorption.

Electrostatic self-assembly is a key aspect of this technology. It allows a uniform formation of multiple, nanometer-thick layers of material into functional ultrathin films, and recent improvements allow the formation of much thicker films and bulk materials. NanoSonic has created a library of similar self-assembled materials, many based on electrostatic self-assembly processing, and has demonstrated the synthesis of more than 2000 individual material layers.

Products

Coatings

NanoSonic offers two eco-friendly HybridShield® coatings: Anticorrosion Coating and Icephobic. HybridShield® Anticorrosion Coating is a single component protective material designed to protect marine, automotive, aerospace, shelter, and communication structures from harsh corrosive environments. In tests, metallic surfaces protected by HybridShield® endured more than 12 months of continuous beach exposure and 5 months of continuous salt fog exposure without signs of corrosion, and exhibited almost no change in color and gloss. All liquid coatings are sold in gallon and quart sizes, at prices ranging from $100-300 per gallon.

HybridShield® Icephobic coating provides higher durability, lower ice adhesion, and reduced ice accretion than competing passive anti-icing protection technologies, according to the company. This material is a two-part fluidic resin with more environmental and mechanical flexibility than competitors, with tailored cure kinetics to ensure easier application with the varied air sprayers found in the industrial coating environment.

Devices

NanoSonic's EKGear Patch monitors EKG signals without using gels or adhesives. It is made of NanoSonic's metal cloth, an electrically conductive cloth that detects the electrical potential that drives myocardial contraction. EKGear materials must be connected or integrated into projects using conductive epoxy, alligator clips, or rivets of conductive materials.

NanoSonic also sells two unique metal rubber products that combine the high electrical conductivity of metals with the stretching capabilities of elastomers. Self-assembly processing allows the simultaneous modification of both conductivity and modulus (stretchability) during manufacturing.

NanoSonic has developed two related products from metal rubber materials: metal rubber electrodes and sensors. Metal Rubber has been demonstrated in a wide range of applications: large mechanical deformation electrodes, mechanically flexible electrical interconnects, and lightweight, durable, conformal electromagnetic shielding. Both products feature malleable metal rubber electrodes that feature a glass transition temperature (temperature at which an epoxy transforms from hard to rubbery) of -60 °C. They have slightly different shapes, and are designed for different applications. The company sells metal rubber electrodes in packs of five 1.5" x 0.5" strips, for $500. Sensors also come in packs of five strips for $500.

Materials

NanoSonic also sells advanced materials directly. Metal rubber sheets are a highly flexible and electrically conductive elastomer, which can be stretched to 1,000 percent of its original shape while staying conductive. The sheets carry data and electricity, and have multiple applications, including power cables, conductors, fabrics, and carbon nanotubes.

Metal rubber addresses a key weakness of carbon nanotubes: once they are deformed, they can lose physical and chemical properties. Making them more flexible—or pairing them with a flexible material like metal rubber—could lead to significant advances in nanotechnology. Metal Rubber sheets are sold in two sizes: 6" x 6" ($1,000) and 12" x 12" ($2,000) sheets.

NanoSonic also sells a fire protection sheet, the HybridShield® Thermal Array. This is a fiberglass sheet that gives extreme fire protection to underlying materials. It is a conformal, highly flexible boundary between firefighters and fire threats that is extremely flame resistant and stable at high temperatures. The company also claims that it provides higher temperature resistance, negligible water absorption, improved impact protection, minimal smoke toxicity, and enhanced flexibility relative to state-of-the-art insulative spacers and energy absorbing materials.

The company anticipates that the HybridShield® Thermal Array will be used for flame/heat protective clothing (firefighting suits in particular), equipment, structures, and vehicles, and has partnered with Shelby Specialty

Gloves to create the next generation of firefighting gloves. The new Thermal Array gloves allow for much more precise movement than today's bulky leather gloves. The Thermal Array is sold in single ($135) and double ($270) sided arrays.

Current Projects

Beyond the existing products described above, NanoSonic is working on projects which it believes will reach the market in the near term. One such project is a new coating for highway barriers, being developed in collaboration with the Federal Highway Administration. When a car collides with highway barriers, the collision generates friction which can roll the car. NanoSonic is developing a coating to be sprayed onto highway barriers that will lessen friction with the aim of reducing rollovers. Tests have been encouraging, although the project is still in development.

Future Products/ Projects

NanoSonic is also currently working to develop aerosol can versions of its HybridShield® Anticorrosion and Icephobic coatings, which the company expects to be available soon, along with Scorpion Skin: a lightweight, conductive, durable, nonwoven polymer matrix resin.

NanoSonic also continues to work on applications related to fire safety. It is developing a new product called HybridShield® CeaseFire—a flame retardant and blast resistant spray. A recent test conducted with the Blacksburg, VA, fire department was very positive. The right side of a derelict building's attic and roof was treated with about 110 gallons of CeaseFire. The treated side did not ignite despite the introduction of additional fuel. It is worth noting that this product has little-to-no toxic byproducts.

Finally, NanoSonic has also been working on optical fiber cables. Many local devices—computers, displays, local area networks—can take advantage of the capacity of an already installed optical fiber network, but need to be connected to it through high-speed links. Standard glass optical fiber jumpers can be used for these links, but they are not cheap or easy to install. With support from DoE, NanoSonic, Inc. has been developing low cost, high performance polymer optical cabling that supports high-speed data over the short distances from the optical fiber backbone to local devices and networks. The fibers are manufactured using the company's patented electrostatic self-assembly process.

Awards

NanoSonic has been recognized by the scientific community, and is the recipient of several notable awards. It was named to the Nano 50—NASA's list of the 50 most impactful nanotechnologies, products, and innovators for its metal rubber fabric technology. The company was also named to the R&D 100 in 2007 and 2011, for metal rubber and fire/blast resistant spray, respectively. Other awards include the Top Small Business Award in VA, a Top 5 Small Business Award at DARPATech, and a Top 13 Nanostructured Products at NASA.

Business Model

NanoSonic's business model is unusual. While most revenues are still derived either from SBIR/STTR awards or from sales of products and services to businesses or to government agencies, it is also now entering direct to consumer sales, for example its glove for firefighters (developed in partnership with a larger company, Shelby Inc.—see below). And NanoSonic also offers both raw materials (sheets of specialized fabric, or coatings) as well as final products such as the glove.

The company's main customers are government agencies, large aerospace, electronics, and biologics companies, and revenues range from $1-$5 million annually. While the company has developed a wide range of technologies with SBIR and STTR funding, and these have constituted a significant amount of revenues to date, NanoSonic is now moving from R&D through product development into manufacturing, and Dr. Lalli anticipates that the balance will tilt further in coming years.

Nanosomic is still focused primarily on R&D—almost all of the current employees are involved in research. However, the company is also reaching out to find new distribution channels, beyond the existing partnership with Shelby. Two additional distribution partnerships are pending as of August 2015, according to Dr. Lalli.

The company is strongly growth oriented. It owns a building with 30,000 square feet of space and with considerable room to expand. It is a "green building," certified by LEED and MAS, and featuring a wall of solar wall panels and other earth-friendly technologies. The facility houses a 10,000 sq. ft. process scale-up and manufacturing lab, and a 10,000 square foot R&D lab. Another 100,000 square foot building is on the drawing board for the facility, to be used for manufacturing. Nanosomic has also always had ambitions to become an international company.

SBIR/STTR

NanoSonic has received 281 SBIR/STTR awards, 243 SBIR and 38 STTR. (206 were Phase I and 75 Phase II). 185 awards have come from DoD and 48 from NASA.

STTR

Dr. Lalli observed that five years ago, she would have wanted to see STTR folded into the SBIR program, in large part because managing ITAR restrictions in the context of a partnership with a research institution was often extremely challenging.

More NanoSonic had found that the process has moved more smoothly, and while there was a clear tension between academic interests in publishing and company needs for confidentiality, this could be addressed effectively with the right partner.

Today, NanoSonic is a very strong supporter of the STTR program, Dr. Lalli said. The company found a formal agreement to use university equipment to be very helpful, and that the program also helped NanoSonic reach out to cutting edge researchers, and gain access to high quality graduate students.

NanoSonic now has good relationships with at least eight universities. Working with other Virginia schools has been especially fruitful—NanoSonic for a long time avoided partnering with Virginia Tech to avoid conflict of interest issues. Other effective partnerships have been formed with Colorado State University, the Naval Postgrad School, and the University of Arizona, according to Dr. Lalli.

Dr. Lalli said that she did not see STTR as presenting more difficulties than other contracts in terms of partners meeting their deliverables. She observed that in both cases, it would be important to figure out the reason for a failure, and to ask the partner for an alternative solution. Sometimes the problem being addressed was just hard, or there were differences of opinion on what needs to be delivered.

NanoSonic always drives the partnership, according to Dr. Lalli. A typical partnership might involve making the materials at the company, with the university providing technical help in measuring performance. For example, in STTR programs with Colorado State University, the partner there is an expert in the measurement of radiation-resistant materials measurement, and also has the necessary equipment in the university lab. He provides evaluations that validate NanoSonic claims, and thus helps the company to improve the material.

Dr. Lalli did however note that the need to deal with ITAR was very challenging. Most SBIR topics from DoD and NASA require this, and NanoSonic is now working to improve its capacity to deal with ITAR-related issues.

Recommendations

Dr. Lalli said that that biggest issue with the program for her company was the lack of clear specifications from DoD for new materials. Simply writing a topic was not enough to ensure that if the material was successful it would have a market within DoD, and she recommended that DoD develop improved procedures for closing the gap between topics and specifications, especially for materials.

PHYSICAL SCIENCES, INC.[34]

Physical Science Inc. (PSI) is a private company founded in 1973 by Robert Weiss, Kurt Wray, Michael Finson, George Caledonia, and other colleagues from the Avco-Everett Research Laboratory. The company is an engineering research and development company, focusing on the application of emerging sciences to the solution of engineering problems for its customers. PSI is headquartered in Andover, Massachusetts, and has approximately 180 employees and annual revenues of over $40 million.[35] Dr. Green has been employed at PSI for 39 years, starting there as a researcher after completing his PhD in chemistry at MIT.

Initially focused on laser and optics-based sensing applications and computer modeling in the aerospace and defense industries, energy sector, and the environment, PSI has over time applied its core expertise to a wide set of technological applications, and in so doing broadened its technical capabilities to include chemicals, materials, and signal processing. By focusing on technological specialties too small to attract major investment from DoD primes contractors and too mission-driven to excite competition from university laboratories, PSI has a solid reputation helping government agencies and private-sector clients across a broad range of technologies, according to Dr. Green. PSI's principal customer is DoD, and its needs for sensing and monitoring technologies has driven the direction and development of PSI's capabilities.

The company is organized into two R&D divisions, Applied Sciences and Defense Systems, and three subsidiary companies, Research Support Instruments, Inc., Q-Peak, Inc., and Faraday Technology, Inc. SBIR/STTR is an important source of funding, especially in developing new competencies, and starting in 1983, PSI has received a total of $284 million in SBIR/STTR funding, while its subsidiaries received $54 million. However, as Dr. Green points out, PSI has always served an array of markets and SBIR/STTR funding has never been more than 50 percent of annual revenues.

---

[34]Primary sources for this case study are the interview with Dr. David Green, CEO September 2 2015, and a review of the Physical Sciences, Inc. website (http://www.psicorp.com) and related company documents.

[35]David Woolf, *et. al,* "High-temperature Selective Emitter for Thermophotovoltaic Energy Conversion," November 12-14, 2014, http://www.psicorp.com/sites/psicorp.com/files/articles/VG14-148-final.pdf, 1.

At its headquarters in Andover, MA, PSI operates over 68,000 square feet of general office, laboratory, and prototyping space. PSI has two satellite offices in Haverhill, MA and Pleasanton, CA. The 6,000 square feet Haverhill facilities perform composites fabrication and laser machining operations and act as a staging area for various experimental activities. The smaller 2,800 square feet Pleasanton, CA facilities focus on nonlinear optics and laser-based chemical sensing. Subsidiaries operate facilities in Maryland, Massachusetts, New Jersey, and Ohio.

Dr. Green noted that a core of 10-20 people have been at PSI for 20 years or more. They understand DoD, NASA, and DoE agency needs. So PSI offers continuity, a deep understanding of the agency mission, and can as a result guide technology development toward meeting agency goals. This is a quite different model than companies seeking to commercialize a single technology, and provides quite different kinds of support to the agencies.

Technology

PSI, since its founding in 1973, has built on its core capacity applying lasers and optics technologies to sensing applications. In the 1980s, with SBIR support, PSI expanded into medical imaging and imaging chemically reacting flows. In the 1990s, PSI extended further into research on materials (especially chemical sensors), batteries, and tunable diode lasers.

Chemistry. PSI works in three broad and interrelated areas of chemistry: energetic materials research (explosives), advanced fuels, and coatings.

Laser-based Sensing. PSI lasers research focuses on three areas: biological structure, physical measurement, and laser spectroscopy. Using optical coherence tomography (OCT), PSI has developed technologies that can capture visually both tissue morphology and function. Based on laser distance and ranging technology (LIDAR), PSI can measure remotely a broad range of the physical and chemical properties of a target and the atmosphere. Finally, with tunable diode laser absorption technology, PSI is developing low-cost, high-volume applications such as natural gas leak detection and greenhouse gas monitoring.

Materials. To support research in energy and sensing applications, PSI developed deep competencies in material science. Initially aligned with its work on lasers, PSI expanded into other materials applications in radio sensing such as chaff manufacture to reduce or distort reflected images. PSI has also developed high temperature ceramics for leading edges and combustors in hypersonic flight and high density energy storage for next generation battery technology.

Optics. PSI has worked in optics since its founding, and as a result has developed technical capabilities in a wide range of areas, including integrated optics, photonics, and non-linear optical materials for gas sensing, field sensing, optical communications, interferometry, industrial process control and non-destructive structural evaluation. Current projects include new imagers, spectrometers, and sensors using digital micromirror device (DMD) technology to increase data rates, improve ruggedness, and reduce the overall size and cost. PSI is also developing materials for applications requiring engineered optical properties for absorption, reflection, and emission at any wavelength.

Passive Sensing. Sensing technology is another longtime core competence of PSI. Current areas of research include magnetometry for measurement of local magnetic fields by drones, surface contamination for detecting environmental chemical agents (explosive or industrial waste), hyperspectral imaging for sensing chemical residues on remote surfaces, and low cost acoustic sensors for determining right-of-way encroachment and excavation activity near a pipeline.

Signal Processing. PSI's work on sensors has also led the company into signal processing. For example, PSI has developed the capability to simulate post-intercept radar scenes with thousands of debris objects. Similarly, the company has a strong portfolio of sonar signal processing analysis models and simulations intended to enhance sonar performance against background noise, clutter, and reverberation.

Products

While PSI is not a manufacturing company and has no plans to become one, its technology does transition into products. Typically, if these have larger scale potential they are licensed to bigger companies for market deployment, while PSI itself may manufacture products that are short run or otherwise low volume.

On its web site, PSI provides a list of 20 products. Some have been licensed for production to other companies, and some are produced in short runs by PSI.

Commercialization: Subsidiaries, Spin-offs, and Licensing

When PSI sees commercial potential in a technology, senior management evaluates the opportunity to determine how to address the opportunity. PSI subsidiaries tend to replicate the R&D culture of the parent company (publication in peer-reviewed journals, use of SBIR funding), to focus on a limited (but stable) commercial opportunity, and to perform prototyping and low volume manufacturing. Spin-outs typically depend on venture backing and incorporate business models targeting larger commercial markets with need for product development, manufacturing, logistics, and sales and marketing.

Subsidiaries

Since 1990, PSI has acquired four wholly owned subsidiary companies. Three continue to operate: Research Support Instruments, Inc. (RSI), Q-Peak, Inc., and Faraday Technology, Inc., while the fourth was sold and now operates as part of a larger company.

Research Support Instruments, Inc.

Founded in 1976, Research Support Instruments, Inc. (RSI) was acquired by PSI in the early 1990s to provide PSI with the capacity to deliver hardware for spacecraft discovery missions as well as on-site engineering support to clients in the DC metropolitan area. The company provides services that enable research and development, systems engineering, instrument test and calibration, and experiment support. It operates offices in Lanham, MD; Princeton, NJ; and at the Naval Research Laboratories (NRL) in Washington, DC. RSI has had some success generating SBIR/STTR funding. Since its founding, RSI has received 44 SBIR/STTR awards, worth $7.8 million. Twelve percent by value have been STTR awards.[36]

Q-Peak, Inc.

PSI acquired Q-Peak in 2001. From its offices in Bedford, MA, the company performs contract research and development in the fields of solid state lasers, nonlinear optics and related technologies. Customers include both the U.S. government and private corporations, especially the aerospace primes. Q-Peak can also produce low volume runs of various devices and systems. Finally, Q-Peak also manufactures a set of products based on diode-pumped, solid state lasers. These standardized, field-proven components can be integrated to provide a broad range of custom functionality. Q-Peak has also had substantial success in acquiring SBIR/STTR funding, having received 110 SBIR/STTR awards, worth $29.4 million. Eight percent by value have been STTR awards.[37]

---

[36]"PSI'S CORPORATE HISTORY," http://psicorp.com/our-company/history; "Excellent Technical Support," http://www.rsimd.com/; "Research Support Instruments, Inc." https://www.sbir.gov/sbirsearch/detail/292228.

[37]"Q-PEAK, INCORPORATED." https://www.sbir.gov/sbirsearch/detail/284118; "Research and Development: Overview," http://www.qpeak.com/Research/roverview.shtml, "Products: Overview," http://www.qpeak.com/Products/products.shtml.

Faraday Technology, Inc.

Faraday Technology, Inc. provides government and commercial clients with R&D services related to electrochemical engineering development running from bench prototype systems through pilot or pre-production levels. By varying the waveform of the applied voltages and currents, the anode/cathode spacing, the anode design, and degree of mixing within a Faraday cell, company technicians can control the electrochemical deposition rates of various atoms. In addition to engineering services, Faraday also markets rectification equipment and effluent decontamination reactor hardware. Faraday Technology has had success generating SBIR/STTR funding, receiving 90 SBIR/STTR awards, worth $21.0 million. Eleven percent by value have been STTR. Faraday also won an R&D 100 Award in 2011 for its work depositing Mn-Co coating on interconnects in solid oxide fuel cells.[38]

Spin-Outs

In addition to establishing subsidiaries, PSI has also spun out technologies into new companies. Typically, these technologies have presented the opportunity for selling products to mass markets. Although PSI may take an equity stake in the company, most of the funding comes from the venture community. The company's record with spin-outs has been mixed.

Confluent Photonics was founded in 2000 to commercialize components for use in Dense Wavelength Division Multiplexing ("DWDM"). Confluent received $14 million in two rounds of venture funding in 2001 and 2003. In 2006, it was acquired by Auxora.[39] Another medical instrumentation firm failed when it couldn't raise a C round to complete clinical trials to gain FDA approval.

Dr. Green said that IP and staff usually go with the spin-out. None of the spin-outs have been highly successful, and many of the staff have returned to PSI. One spin-out still exists, having been sold three times. Spinouts are

---

[38]"The Company," "The Technology," http://www.faradaytechnology.com/; "FARADAY TECHNOLOGY, INC." https://www.sbir.gov/sbirsearch/detail/164726; "Faraday Wins R&D Magazine's R&D 100 Award," http://www.faradaytechnology.com/PDF%20files/FT%20R&D%20100%20Press%20Release.pdf.

[39]"Confluent Photonics Raises $11 Million Series A Financing," January 10, 2001, http://www.prnewswire.com/news-releases/confluent-photonics-raises-11-million-series-a-financing-from-innocal-venture-capital-rustic-canyon-ventures-cit-venture-capital-and-invesco-private-capital-71002827.html; "Confluent Photonics Raises $3 Million in Second Round Financing," September 11, 2003, http://www.prnewswire.com/news-releases/confluent-photonics-raises-3-million-in-second-round-financing-71066127.html; "Auxora Acquires Confluent Photonics," March 6, 2006, http://www.auxora.com/doce/news-detail-26.html.

however in the end are in the hands of the investors who buy control. In some cases, that can be invaluable where they provide good market insight. However, in many cases the technology takes too long to mature, and investors take the new company in the wrong direction. A good recent example would be 3-D cinema—the company's technology was in that case transferred to outside group which lacked the technical capacity to execute the project effectively.

Licensing

PSI has licensed significant amounts of technology. Perhaps the most successful example is the Remote Methane Leak Detector (RMLD™), a laser sensor used worldwide to detect natural gas leaks. PSI began RMLD™ development in 1999, initially funded by EPA Phase I and Phase II SBIR grants and subsequently with funding from the Department of Energy and industry partners. The eventual product is a hand held device that can detect methane from outside the plume. According to Dr. Green, PSI developed the product all the way through to a pre-production prototype. It worked collaboratively throughout the development with an industrial partner as well as national gas distribution companies.

Four years work resulted in a prototype. PSI licensed the RMLD™ technology to Heath Consultants, Incorporated on an exclusive basis in 2003, and renewed the license for another ten years in 2013. Heath released RMLD™ commercially in 2005 and has since sold over 3,000 units worldwide at about $17,000 each, generating revenues of approximately $50 million and PSI royalties of $2 million. The detector has spawned its own cluster of jobs through companies using the detector, which Dr. Green estimates at more than 3,000 employees along with commensurate tax revenues. The product team received a 2005 R&D 100 Award. In 2006, PSI received a Tibbetts Award.[40]

According to the PSI web site, PSI licensing income recently exceeded $1 million annually following the successful commercialization of its ophthalmic technologies.

Patents and Other Intellectual Property

PSI is the assignee for 70 patents over the period 1987 to 2015. Five patents have multiple assignees reflecting R&D collaboration between PSI and other organizations. They were Faraday, Incorporated; American Air Liquide,

[40]"Tibbetts Award Winners," http://www.sbtc.org/tibbettswinners/; "Detecting gas leaks from a distance," August 31, 2005, http://www.rdmag.com/award-winners/2005/08/detecting-gas-leaks-distance.

the General Hospital Corporation, and Alliant Techsystems. Almost half (32) of PSI patent portfolio has been published in the past 5 years which suggests that PSI's patent strategy has changed.

Partnerships

PSI maintains research relationships with a broad range of university, government, and corporate R&D organizations. For example, PSI has recently successfully licensed technology for ophthalmic instrumentation to both an incumbent and two start-ups. The technology was developed in partnership with scientists, engineers, and clinicians from organizations like the Army Medical Research Branch, the Air Force Research Lab, the Massachusetts Eye and Ear Infirmary, MIT, the University of Texas at Austin, Massachusetts General Hospital, Boston Medical Center, and Brigham and Women's Hospital.[41]

Revenues

PSI generates over $40 million annually in revenues, down slightly from its peak in the late 2000s. The company has received extensive support from SBIR/STTR funding. It also generates revenue from engineering service contracts, product sales from its subsidiaries, technology licensing, and to a lesser extent spin-outs.[42] PSI reports its revenue breakdown for FY2010 as that listed in Table E-7 (including subsidiaries).[43]

SBIR/STTR

Between 1983 and 2015, SBIR/STTR funded 1,108 projects with PSI: $63 million in SBIR Phase I, $190 million in SBIR Phase II, $8.0 million in STTR Phase I, and $23.4 million in STTR Phase II funding. PSI's subsidiaries have also benefited from SBIR/STTR, receiving an additional 244 awards worth $58 million.[44]

---

[41]Dan Hammer, *et. al.* "Biomedical Optics Instrumentation," September 15, 2010, http://www.psicorp.com/pdf/library/VG10-182.pdf, p. 7.
[42]Dan Hammer, "Biomedical Optics Instrumentation,"http://www.psicorp.com/pdf/library/VG10-182.pdf, 1; Woolf, "High-temperature Selective Emitter," http://www.psicorp.com/sites/psicorp.com/files/articles/VG14-148-final.pdf, p. 1.
[43]David Woolf, *et. al,* "High-temperature Selective Emitter," http://www.psicorp.com/sites/psicorp.com/files/articles/VG14-148-final.pdf, p. 1.
[44]"PHYSICAL SCIENCES, INC." https://www.sbir.gov/sbirsearch/detail/273626; National Research Council, *An Assessment of the SBIR Program*, Washington, DC: The National Academies Press, 2008, http://www.nap.edu/catalog/11989.html, p. 268.

TABLE E-7 Physical Sciences, Inc.'s Revenue Breakdown, FY2010

Percent of FY2010 Revenue	Source of Funding
60	Applied research and development for U.S. government agencies
20	Components, systems, and instrumentation for industry and government sales
15	Product development and commercialization for government and industrial customers
5	Development of pre-production manufacturing processes
2	Licensing fees from strategic partners and spin-outs for high-volume commercial markets

SOURCE: Physical Sciences, Inc.

Of the 93 SBIR/STTR projects awarded to PSI in 2013 and 2014, 61 percent (57 projects) were funded by DoD, 17 percent by NIH, and 12 percent by DoE. The remaining 10 percent were funded by the Department of Agriculture, the EPA, the Department of Homeland Security, and the National Aeronautics and Space Agency. Over the more than 30 years that PSI has received SBIR/STTR funding, STTR awards account for just under ten percent by value.

Both PSI and the SBIR/STTR programs have evolved over time. Initially, the company was focused on basic and near basic research. Today is it working on applied research and then applications and commercialization. Dr. Green said that the company was already evolving towards a more pronounced focus on commercialization before more recent changes in the SBIR/STTR program in the same direction.

Today, PSI is a strong supporter of the program's shift away from research-only projects. The company no longer just looks for projects that it can win—managers want to know where the technology will be used, and they want to see effective commercialization, according to Dr. Green. Before staff write a Phase I proposal, the company has to have a commercialization plan—it is part of the bid decision for PSI. And while PSI still sees itself as a research house, it is now focused much more closely on applications for that research.

Dr. Green said that successful commercialization of SBIR technologies—especially from DoD and NASA SBIR/STTR projects—required that the company find multiple markets—simply relying on direct agency sales was not sufficient. Thus while PSI's work with NASA had led to a number of commercial successes, these had not been through direct sales to NASA. Diagnostic tools developed for NASA, for example, are now used in the automotive industry. Similarly, PSI is currently building an aviation fuel quality monitor for Navy for aircraft carriers. Orders for these monitors come once every 3 years, so that business alone cannot sustain a company.

STTR

Dr. Green said that he was a strong supporter of the STTR concept. However, while STTR provides funding for the research institution, industry has to be the bridge that transitions technology out of academia. STTR cannot just be pass through funding to the RI. He believes that STTR encourages each partner to work to their strength: the RI does research and education, and the industry partner does commercialization, and this structure is perfect for technology transition.

Dr. Green observed that PSI had spent more than $9 million on contracts with RIs since 2009. Most of that has been through SBIR/STTR (though there have been some other contracts). In one six year period, PSI funded 53 different universities. The company watches the scientific literature to identify possible partners, focusing on faculty who are making cutting edge advances that can meet the needs of PSI's customers. It is rare that a professor says they are not interested in collaboration.

PSI has had a number of successful STTR projects. One focused on imaging of the retina, and stretched over several STTR awards, starting with NIH support. NIH wanted technology to detect macular degeneration earlier, and the technology might also help detect eye diseases in premature infants.

The objective of the project was to resolve to very fine level the vasculature at the back of the eye at the surface and in depth. That allows clinicians to understand the dynamics of the back of the eye using optics only.

PSI had worked on the project with a number of high quality academic partners in the Boston area, including Children's Hospital. Working closely with top researchers, seeing their challenges and identifying tools to resolve them, before working together on clinical trials and further refinement of the tool is highly satisfying for PSI researchers. Publishing academic papers jointly was also important—it allowed new ideas to emerge from the scientific audience, and often stimulated possible new applications for the tool. Dr. Green thus saw the project as creating a powerful virtuous circle: PSI staff are instrument builders, not clinicians, but the company's work helps the clinicians do things they could never have done otherwise. That in turn created more publications and more recognition for the project, and ultimately patents that were filed jointly with RIs such as Children's Hospital.

The product of the STTR-funded research has now been licensed to major medical device companies, as it is not realistic for a company the size of PSI to fund clinical trials. Dr. Green said that PSI now receives modest royalties, as the device companies sell the product. Over the past 7 years, 15,000 units have been sold, generating approximately $1 billion in revenues. Perhaps more important, tens of millions of patients have been tested using this technology, improving health for everyone.

While Dr. Green supports STTR, he said that it was not clear that it added substantial value beyond SBIR. PSI works with RIs through both programs, and finds that RIs are brought into projects because they are needed.

There is in his view no difference in the company's management of SBIR and STTR programs. All subcontractors need to be managed, which is especially hard to do in the short timeframe of a Phase I award. Universities may even be a little easier to manage than collaborations or subcontracts with large technology companies.

Recommendations

Dr. Green said that overall the review process at the agencies was high quality, particularly at DoE. It often provided many different technical perspectives, which was valuable. Commercial review was probably not as insightful, but noone can perfectly see the path to commercialization. Efforts have been made to improve commercial review, and DoE in particular has tried to raise awareness and improve quality. He suggested that agencies should seek expert advice on commercialization, which was now widely available in the private sector. Reauthorization has resulted in more reporting and a lot more structure. The amount of effort required to submit a proposal has more or less doubled even for a highly experienced company like PSI. This represents a major barrier to entry into the program: Dr. Green noted that the grants.gov SBIR/STTR instructions are 200 pages long, which may in part explain why the number of proposals is falling. Every SBIR/STTR proposal requires that PSI uploads 10-30 different sections. One has to be very internet savvy and very persistent.

Dealing with government has in general become much harder. Now numerous forms and statements are required related to fraud and abuse: all proposals now require that the company has support for every piece of equipment it plans to buy, and provide support to show that it is actually paying everyone that it plans to pay.

The agencies need to look again at this, to find ways to simplify the process substantially, to limit the amount of paperwork involved in an application. Everyone should have a fair shot, and that is not really the case today. PSI has a fully trained technical publications department to do submissions and it still takes them significant time and effort. It is important that the program remain fully merit-based, to ensure that the best solutions find their way to the market.

STRATATECH CORPORATION[45]

Stratatech Corporation is a private company founded in 2000 by B. Lynn Allen-Hoffmann. The company is developing novel skin regeneration and repair products for therapeutic use, drawing on what Dr. Allen-Hoffmann described as a serendipitous discovery in her lab at the University of Wisconsin that offered entirely new technical opportunities in cell-based therapy for human skin replacement and treatment of complex skin defects.

Working together with the University of Wisconsin and the Wisconsin Advanced Research Foundation (WARF), Dr. Allen-Hoffmann used an STTR award to begin the transition from university lab to the private sector. In conjunction with WARF, she determined that a small private biotechnology company was the appropriate legal structure to house the work, and provided access to the SBIR program.

Stratatech started operations in a small space provided by Mirus Corporation, another small spin-off of the university located in the University Research Park in Madison, Wisconsin. The company soon started to attract angel funding, which Dr. Allen-Hoffmann attributes to the understandable nature of the technology for skin replacement. While business advisors recommended that she avoid applying for grants due to the lengthy time required to generate the application and administer the grants if awarded, Dr. Allen-Hoffmann decided that the best path to funding lay through the SBIR/STTR programs.

Based on the discovery of a human keratinocyte cell line, NIKS® cells, that produces tissue nearly identical to human skin, Stratatech has used the cells as a platform technology for the development of a pipeline of cell-based products. Stratatech is developing StrataGraft® as it's flagship product based on the core technology. StrataGraft® is a living skin-tissue therapeutic for the treatment of severe burns and other complex skin defects, the use of which may reduce or possibly avoid the need for painful skin harvest and transplantation (autografting). The ExpressGraft™ lineage is comprised of skin tissues that have been genetically enhanced to encourage wound closure by providing elevated levels of human wound healing or antimicrobial factors that may be underrepresented in some wound environments. Both the core technology, Stratagraft®, and the world's first genetically enhanced human skin, Expressgraft™, are being evaluated in late-stage and early-stage clinical trials, respectively. The late-stage clinical development supporting the StrataGraft® product is in part funded by a $247 million contract with Biomedical Advanced

---

[45]Primary sources for this case study are the interview with Barbara-Allen Hoffmann conducted on April 9, 2015, and a review of the Stratatech web site (http://www.stratatechcorp.com) and related company documents.

Research and Development Authority (BARDA) awarded in September 2015. Results to date have supported the safety and initial efficacy of the company's flagship product, StrataGraft®.

By late 2013, Stratatech had 38 full-time employees and expected to add 10 to 20 additional employees over the next 5 years. Currently, the company has approximately 50 full-time employees. It has research relationships with various universities and research institutions including the University of Wisconsin-Madison, Wake Forest University, The Arizona Burn Center, the U.S. Army Institute of Surgical Research, Harvard Medical School, and an unnamed Fortune 200 consumer products company.[46] However, even with a large support contract in hand from HHS/BARDA and continuing support from NIH, funding for later stage clinical trials and manufacturing infrastructure remains an ongoing challenge. Dr. Allen-Hoffmann observed that there had been no new products available for burn patients in decades, in large part because the market was perceived as too small to interest large biomedical companies. In 2012, StrataGraft® received orphan drug designation from the FDA to expedite treatment for severely burned patients.

Technology

Unlike other cultured human cell lines, the NIKS® progenitor line at the heart of Stratatech's core technology is a consistent source of pathogen-free, non-tumor-producing, long-lived adult keratinocyte progenitor cells. Keratinocytes are the cells that make up approximately 90 percent of the outer layer of human skin known as the epidermis. The value of the NIKS® cell line lies in its ability to regenerate the epidermal component within a fully stratified human skin tissue. The resulting multi-layered tissue has the physical strength and biological characteristics of intact human skin. When handled appropriately, this cell line grows new human skin and—as important—ceases growth when these cells abut neighboring mature skin cells. The NIKS® cell line can be utilized indefinitely to produce cultured skin, avoiding the costly need to recreate and requalify new cell lines that restricts other technologies.

Having a well characterized, pathogen-free, continuous source of epidermal progenitor cells serves as a foundation for the company's products and allows Stratatech to pursue strategies to improve the cell line's performance genetically. Stratatech is introducing new genetic characteristics without using viral vectors or other delivery technologies that could impart unwanted safety risks to the transgenic tissue and, most importantly, the patient. This approach

---

[46] "Company Profile: Stratatech," http://inwisconsin.com/insource-newsletter/Stratatech-company-profile/.

supports the creation of custom cell-based therapeutics with enhanced antimicrobial properties and improved vascular function and that may lead to faster healing.

Products

StrataGraft® and the ExpressGraft™ line of genetically enhanced tissues are the principal products under development from the NIKS® cell line. Currently, the company has created six ExpressGraft™ pipeline products, each genetically augmented to address the underlying pathophysiology of complex skin defects. All pipeline products have been successfully developed from hypothesis to completed cGMP manufactured master cell banks with support from the SBIR/STTR Program.

StrataGraft® Regenerative Skin Tissue

Each year in the United States, about 40,000 hospitalizations occur for burns.[47] At present, patients with severe burns must endure autografting, a procedure requiring the harvest of healthy skin from another part of the body for transplantation to the site of the wound. StrataGraft® tissue has the potential to provide a safe, effective, and less painful alternative that avoids the creation of donor site wounds.

StrataGraft® skin tissue is a cellular therapeutic for use as a treatment for severe burns and other complex skin defects. It mimics natural human skin, with both dermal and fully-differentiated epidermal layers. StrataGraft® skin tissue is easily sutured to a wound bed, provides barrier function, and is anticipated to serve as a source of factors promoting the natural skin regeneration process.

In October 2014, StrataGraft® completed a Phase Ib clinical trial in patients with deep partial-thickness burns. By 90 days after treatment, 27 of 28 patients achieved complete wound closure after a single application of StrataGraft® tissue. By this time, no StrataGraft® DNA was detectable, confirming regeneration of the patients' own skin.

If successful, StrataGraft® could revolutionize treatment for burns by providing an alternative to autografting and its associated donor site pain, infection risk, and possible poor cosmetic outcome. These advantages may lead to shorter hospital stays and reduced after-care costs.

---

[47]American Burn Association, "Burn Incidence and Treatment in the United States: 2013 Fact Sheet," http://www.ameriburn.org/resources_factsheet.php.

ExpressGraft™ Genetically Enhanced Regenerated Tissue

Approximately 50 million surgeries occur annually in the United States, each requiring some form of wound closure.[48] Stratatech is developing genetically enhanced tissues that produce elevated levels of natural wound healing and antimicrobial factors. Delivered as skin grafts, ExpressGrafts™ create a controlled wound microenvironment in which to fight infection or promote vascularization while accelerating healing.

In one ExpressGraft™ product, the NIKS® cell line has been genetically modified to produce higher levels of cathelicidin, a peptide with antimicrobial properties that plays an active role in wound healing. These tissues produce 140-fold greater levels of cathelicidin protein *in vitro*, and in an *in vivo* animal wound model showed a 100-fold reduction in the presence of a multidrug-resistant strain of *Acinetobacter baumannii.*

Clinical development of ExpressGraft™ will start in 2015 with a Phase I/II trial focused on non-healing diabetic foot ulcers. An IND was submitted to the FDA in spring 2015 and received clearance from CBER to enter a first-in-human safety trial. Dr. Allen-Hoffmann observed that this project has been supported from "hypothesis to translation into the clinic" by NIH through the STTR and SBIR programs.

StrataTest® Human Skin Research Model

Many of today's animal- and cell-based toxicity testing models are burdened by significant accuracy, reproducibility, cost, and ethical concerns. The European Union, for example, has banned the sale of animal-tested cosmetic and consumer products.

Based on the NIKS® cell line, StrataTest® is a human skin model for *in vitro* consumer product testing, drug discovery and toxicity screening. Like StrataGraft®, StrataTest® tissue is composed of both epidermal and dermal layers, and displays the same physical, chemical and histological characteristics of human skin, enabling better prediction of *in vivo* biological responses than monolayer skin culture technologies.

Dr. Allen-Hoffmann said that while StrataTest® has shown considerable technical promise and the company regularly fields inquiries from larger potential customers, the decision was made to focus efforts on the therapeutic flagship StrataGraft® product and the ExpressGraft™ pipeline of products for the time being.

---

[48]CDC FastStats, http://www.cdc.gov/nchs/fastats/insurg.htm.

Other Potential Products

Other ExpressGraft™ potential products are in the pipeline. Like the cathelicidin-expressing variant of ExpressGraft™, some product candidates produce elevated levels of other naturally-produced human wound healing factors. For example, one proposed product expresses VEGF, a protein that plays a central role in blood vessel growth (angiogenesis). Because many chronic wounds are associated with insufficient tissue oxygenation, boosting local levels of VEGF may improve wound healing and closure. Clinical development will target the need for underserved markets in chronic, non-healing ulcers.

Additional potential products target different classes of skin trauma. For example, by creating tissues that express Interleukin-12 (IL-12), a human anticancer protein, Stratatech hopes to develop a product that surgeons could apply after surgical excision of solid skin tumors. Locally produced IL-12 from the genetically modified tissue could facilitate the patient's own immune surveillance of residual tumor cells remaining after the surgery, reducing the potential for recurrence.

Patents and Other Intellectual Property

Stratatech is the assignee for 20 issued patents listed in Table E-8.

Funding

Stratatech Corporation has received grant support from SBIR (mostly from NIH but also from DoD), other grants from non-SBIR sources, a major contract from HHS's Biomedical Advanced Research and Development Authority (BARDA), and investment from independent investors.

Non-Dilutive Grants

Between 2001 and 2013, SBIR funded 21 projects with Stratatech. From 2001 to 2003, it received Phase I SBIR awards from four NIH centers—National Institute of Arthritis and Musculoskeletal and Skin Diseases, National Institute of General Medical Sciences, National Cancer Institute, and National Institute of Environmental Health Sciences—followed in 2004 by the first Phase II award from NIGMS. Subsequently, Stratatech also received awards from NIDDK and NIA.

STTR grants funded three projects aimed at completing the scientific work that, according to Dr. Allen-Hoffmann, needed to be done within her lab at the University of Wisconsin because that provided access to otherwise unaffordable equipment.

Stratatech has also applied for and received Fast Track awards from NIH. Dr. Allen-Hoffmann observed that these had been especially helpful as they reduced the time from initial idea to clinical trials by many months. One Fast Track provided by NIDDK is supporting Phase I clinical trials for an anti-infective human skin tissue that can be used to treat ulcerated skins from diabetic skin ulcers.

Stratatech has received grants from other sources to support commercialization of its StrataGraft® product. In July 2013, Stratatech received a grant for up to $47.2 million from BARDA. The award supports the preclinical, clinical, regulatory, and technology development activities needed to complete FDA approval for StrataGraft®. Also, the contract funds

TABLE E-8 Stratatech Assigned Patents

Patent Number	Patent	Year
9,163,076	Human skin equivalents expressing exogenous polypetides	2015
8,992,997	Dried and irradiated skin equivalents for ready use	2015
8,808,685	Method of treatment using organotypically cultured skin tissue comprising NIKS® cells that express exogenous HIF-1.alpha.	2014
8,790,636	Human skin equivalents expressing exogenous polypeptides	2014
8,685,463	Dried and irradiated skin equivalents for ready use	2014
8,580,314	Dried and irradiated skin equivalents for ready use	2013
8,092,531	Human skin equivalents expressing exogenous polypeptides	2012
7,988,959	Method of treatment using organotypically cultured skin tissue comprising NIKS® cells that express exogenous HIF-1a	2011
7,955,790	Skin substitutes with improved barrier function	2011
7,915,042	Keratinocytes expressing exogenous angiogenic growth factors	2011
7,888,496	Kit for species specific DNA detection	2011
7,807,148	Organotypically cultured skin tissue comprising NIKS® cells that express exogenous HIF-1a	2010
7,674,291	Human skin equivalents expressing exogenous polypeptides	2010
7,541,188	Skin substitutes and uses thereof	2009
7,501,238	Skin Substitutes for irritancy testing	2009
7,498,167	Keratinocytes expressing exogenous angiogenic growth factors	2009
7,462,448	Species specific DNA detection	2008
7,407,805	Skin substitutes with improved barrier function	2008
6,974,697	Skin substitutes with improved barrier function	2005
6,846,675	Skin substitutes and uses thereof	2005

SOURCE: U.S. Patent and Trademark Office.

manufacturing process development and scale-up to enable large-scale production in case of a mass casualty event.[49] In September 2015 Stratatech received a second BARDA contract through Project BioShield that replaced the first contract. This most recent BARDA contract enables expansion of the company's clinical program to include pediatric patients and aging adults and positions StrataGraft for use under a pre-Emergency Use Authorization, provided the clinical findings support continued development of the product. Importantly, the new BARDA contract included the procurement of StrataGraft by the U.S. government in the event of a mass casualty caused by a natural disaster or an act of terrorism.

In 2010 the Defense Department's Armed Forces Institute of Regenerative Medicine (operating in conjunction with Wake Forest University) funded the proof of concept Phase IIB clinical trial of StrataGraft®. In 2015 Stratatech received approval from the FDA to continue with a Phase 3 clinical trial, based on results from the Phase IIB. The Phase 3 trial will start in early 2016, to be funded by BARDA's Project BioShield.

Equity Funding and Operations

Stratatech has received ongoing support from Wisconsin's angel investor community and from the Wisconsin Advanced Research Foundation. For example, in May 2010 it announced $3.0 million in funding comprised of convertible notes from its current investors.

Strategic Partnership

In 2010 Stratatech entered into a collaborative agreement with a Fortune 200 consumer products company to develop an advanced skin care product. Dr. Allen-Hoffmann said that the objective was to develop the capability to provide testing kits for skin care products. The announced goal was to use extracts from the NIKS cell line to prevent wounds or ulceration by enhancing the resiliency of compromised or susceptible skin.

SBIR/STTR at Stratatech

Dr. Allen-Hoffmann stressed that the SBIR/STTR program at NIH had provided absolutely critical funding for Stratatech. She said that she had no doubt that her company and its associated products would not be in existence

---

[49]Stratatech, "Stratatech Awarded BARDA Contract Valued up to $47.2 Million for Advanced Development of StrataGraft® Skin Tissue for Thermal Burns," Press Release, July 31, 2013, accessed at http://www.stratatechcorp.com/news/20130731.php.

without SBIR/STTR funding. She also observed that the funding was especially important for a woman-owned company: other sources of capital were, in her view, even more inaccessible for a woman-owned small high-tech firm than they were for small high-tech firms in general.

In her view, STTR was particularly important. Some of the initial work—such as work on genetically enhanced tissues—had to be completed in the university lab as necessary equipment was not available at the University Research Park. Once Stratatech was established as a functioning company and the basic research had been completed, other sources of funding became more available.

Today, academic institutions continue to view STTR more favorably than SBIR, particularly with regard to issues related to the allegiance of faculty. University departments take a different view of projects where more of the work and most of the PI's time is committed to the university as opposed to the private sector. Dr. Allen-Hoffmann observed that despite some changes, tenure decision committees were still very conservative about the activities of junior faculty outside academia, and STTR provided an important mechanism for helping to resolve that tension.

Dr. Allen-Hoffmann said that Stratatech had participated in the Fast Track program in the early 2000s when working on developing cell-based ExpressGraft clones. The company feared that the Phase I-Phase II gap would kill the project. Fast Track had worked perfectly from the company's perspective. It had provided a seamless transition from Phase I to Phase II; in her view the company would have lost key people without it. Continuity of staffing remains a key issue for small companies.

Recommendations

Dr. Allen-Hoffmann observed that the SBIR program coordinators at each of the Institutes and Centers played a critical role. Although program officers in general have a strong commitment to the SBIR/STTR program, the SBIR program coordinators possess specific knowledge and can be extremely helpful in guiding investigators. She recommended that small companies make sure that they established contact with the program coordinators.

She also noted that the alignment between topics and awards had changed significantly over the past ten years. During her early years with the program, Dr. Allen-Hoffmann said that she was confident that a strong project would receive consideration and perhaps funding regardless of its connection to a topic described in the Omnibus Solicitation. That has changed over the years, and Stratatech now only applies for awards where there was a clear alignment between the topic and the proposal. In her view this was not a positive development for identifying and supporting innovation.

In addition, Dr. Allen-Hoffmann noted that contracting had become more complex because it was no longer possible to interact routinely with specific financial management officers at NIH. As a result, the advice received started to lack continuity. Continuity is especially important to a small firm trying to budget accurately.

Overall, Dr. Allen-Hoffmann said that she remains truly grateful for the support provided by the NIH SBIR/STTR program and that the technology could not have been developed without that support. The value of this program is immeasurable in helping patients and their families benefit from the world-class research conducted in the United States.

VISTA CLARA, INC.[50]

Vista Clara is a private company founded in 1997 by Dr. David Walsh, a design engineer with experience developing magnetic resonance imaging systems (MRI) in the healthcare industry. Dr. Walsh said that he had been an entrepreneur even before graduate school, and that he had always wanted to own his own company. After completing graduate school, he had started a Vista Clara as a technology consulting company in Tucson, and it had been growing slowly but steadily when he decided to start applying for SBIR funding. The resulting awards allowed the company to develop its core technology (see SBIR/STTR and Vista Clara section below).

Vista Clara develops and manufactures advanced nuclear magnetic resonance (NMR) geophysical instrumentation systems for groundwater, mining, and environmental studies. Vista Clara's NMR instrumentation can operate from the surface, downhole, or in the laboratory, and delivers quantitative imaging of subsurface hydrogeologic structure. The company both sells and rents this equipment, and provides training in its use. Vista Clara also uses its own equipment to perform hydrogeologic field surveys for customers ranging from private land-owners to government agencies and multinational firms.

In 2002, Vista Clara Inc. pivoted from its initial focus on healthcare MRI to applications of NMR to hydrogeology. SBIR funding enabled the company to develop its first NMR based system for groundwater surveying. Although initially expecting to focus primarily on the U.S. market, Vista Clara has found greater market acceptance overseas, principally in China and Australia. Exports are the basis of the company's revenue and profit growth.[51]

---

[50] Primary sources for this case study are the interview with Dr. David Walsh, CEO, August 18, 2015, and a review of the Vista Clara web site (http://www.vista-clara.com) and related company documents.

[51] David Walsh, "Use of Exports to Accelerate Adoption of NMR Geophysical Technology," National Groundwater Association, Theis Conference, November 8-10, 2013, Phoenix, Arizona, https://ngwa.confex.com/ngwa/theis2013/webprogram/Paper9564.html.

Vista Clara is receiving recognition for its work. For example, Elliot Gruenwald, the chief geophysicist for Vista Clara, recently won the J. Clarence Karcher award from the Society of Exploration Geologists for his innovative work on surface NMR.[52]

The company's clients includes various corporate (Rio Tinto, BHP Billiton), university (Rutgers University, Stanford University), government (U.S. Geological Survey, Kansas Geological Survey, Qinghai Geology and Mineral Exploration Bureau, Geoscience Australia) and NGO (Geophysicists without Borders) entities.

Vista Clara currently employs approximately 15 people at its Mulkilteo, Washington headquarters. To serve Asian markets, Vista Clara also maintains a small office in Perth, Australia.

Technology: NMR Hydrogeologic Instrumentation

Water scarcity affects every continent. By 2025, around 1.8 billion people will be living in areas of absolute physical scarcity; two thirds of the world's population will be living under water stress. For many, underground aquifers are an important source of water. However, in most parts of the world, the data required for principled management of these resources is lacking and groundwater aquifers are being depleted.[53]

Vista Clara is developing NMR products and services to measure groundwater. NMR is a physical phenomenon whereby certain elements absorb and re-emit electromagnetic radiation. The sensing using NMR is a two-step process. First, the magnetic spins in a sample are aligned using a magnetic field, and second a radio pulse perturbs the aligned fields. The exact frequency of the pulse depends on the atom to be detected and the strength of the magnetic field.[54]

Conveniently both hydrogen and carbon show this phenomenon. NMR was first applied in geophysics to oil exploration in the 1960s to help develop

---

[52]Rosemary Knight, "J. Clarence Karcher Award for Elliot Grunewald," *The Leading Edge*, January 2015, 15; http://www.tleonline.org/theleadingedge/january_2015?pg=15#pg15.

[53]*Non-renewable Groundwater Resources*, Stephen Foster and Daniel Loucks, eds., Paris: United Nations Educational, 2006), 81; http://unesdoc.unesco.org/images/0014/001469/146997E.pdf; "water & poverty, an issue of life & livelihoods," FAO Water , http://www.fao.org/nr/water/issues/scarcity.html.

[54]Abi Berger, "How Does It Work: Magnetic Resonance Imaging," *British Medical Journal*, January 5, 2002, vol. 324, no. 7328, p. 35, http://www.ncbi.nlm.nih.gov/pmc/articles/PMC1121941/; Allan Newman, "Between a Rock and a Magnetic Field: Geologists Exploit NMR," *Analytical Chemistry*, August, 1991 vol. 63, no. 8, p. 467, http://pubs.acs.org/doi/pdf/10.1021/ac00008a732.

understanding of oil flows through hydrocarbon-bearing rock. NMR instruments designed for the oil industry, however, are generally overengineered for hydrological field studies. The hydrogeologic bore holes are substantially narrower, the physically constants of the targets are different, and the operating temperatures and pressures substantially lower.[55] In a hydrogeologic study, NMR allows the measurement of key hydrological soil characteristics. It can distinguish between bound water that will not flow and unbound water that will. From this, it can also determine the porosity of a soil, a crucial variable in determining flow through that soil.

Initially, Vista Clara developed innovative non-invasive multi-channel (GMR) surface systems designed to enable rapid evaluation of water aquifers without drilling expensive exploratory wells.[56] In the past 8 years Vista Clara emulated the oil industry NMR instrumentation systems for hydrogeologic NMR systems that functioned down bore holes (Javelin) or in laboratories (Corona).

Products and Services

Vista Clara has created a product line that provides different ways of using NMR to evaluate near surface geology (surface-based, small bore holes-based, laboratory-based).

Instrumentation

Vista Clara offers four different instrumentation packages:

GMR. GMR is a surface magnetic resonance sounding systems that allows non-invasive detection and measurement of ground water. GMR uses the earth's magnetic field to align the hydrogen atoms in the water molecules and broadcasts an electromagnetic pulse from surface electrodes to generate an NMR response. Sensors detect the return signal enabling groundwater and soil characterization to a depth of 150 meters without the need for drilling bore holes. Applications include groundwater exploration and well site selection.

---

[55]David Walsh, *et. al.* "A Small-Diameter NMR Logging Tool for Groundwater Investigations," *Groundwater* November-December 2013, vol. 51, no. 6, 914-915, http://www.alphageofisica.com.br/vista-clara/papers/groundwater_javelin_www.alphageofisica.com.br.pdf.

[56]David Walsh, "Multicoil low-field nuclear magnetic resonance detection and imaging apparatus and method," U.S. Patent 8,451,004, May 28, 2013, http://patft.uspto.gov/netacgi/nph-Parser?Sect1=PTO2&Sect2=HITOFF&u=%2Fnetahtml%2FPTO%2Fsearch-adv.htm&r=1&p=1&f=G&l=50&d=PTXT&S1=8451004.PN.&OS=pn/8451004&RS=PN/8451004.

Javelin. Javelin was designed to profile the hydrological characteristics of the geology surrounding a bore hole. Designed for older well fields in which a network of monitoring wells already exists, Javelin is lowered down each bore hole, developing a vertical profile of the hydrological properties for the soils surrounding the bore.

Discus. Discus is a surface technology that enables rapid characterization of surface soils using NMR without the need for sample extraction, porosity calibration, or radioactive sources. Discus can be rapidly moved across a site to develop a two dimensional map of surface soil characteristics. Applications include non-invasive studies of agricultural drainage, roadway compaction, and moisture in building concretes.

Corona. Corona is a portable system for evaluating the hydrological characteristics of soil cores. Using the same technology as a MRI scanner, Corona exposes a sample to a strong magnetic field and then a series of electromagnetic pulses. This system can be used for engineering, geotechnical, or agricultural studies of soil cores. Vista Clara also uses Corona-enabled core studies to calibrate Javelin and GMR/Discus data sets.

Rental and Training

To enable broader adoption of NMR technology, Vista Clara enables customers to rent NMR products for periods ranging from a few days to a few months. To ensure that data is properly captured and analyzed by both rental and first time customers, Vista Clara personnel will travel to provide on-site training.

Field Surveys

Vista Clara will perform custom field surveys for its clients, although according to Dr. Walsh it prefers to train client staff. It will assist in study design, data acquisition, data review and processing, data interpretation, and report preparation.

Markets

Vista Clara sells small numbers of moderately expensive equipment (GMR systems are approximately $200,000 each), so individual sales have a real impact on the company, according to Dr. Walsh.

In general, Vista Clara sees its markets as global. The company has found that demand for its products fluctuates, but not simultaneously in all markets. In China, the company found an effective distributor for geophysical instruments and had two years of growth, but the recent slowdown of the Chinese economy has limited opportunities in that market. Thus the sale of 3 GMR systems in 2013 has been followed this year by the sale of one system. The company is now seeking to develop systems that can be sold at a lower price, in an effort to build the volume of sales and make revenues less volatile.

Governments are the primary end users of the data generated by Vista Clara systems. Projects involving the systems tend to be large scale—for example, a recent project maps the aquifers of western Nebraska. As a result, systems are typically bought by government agencies or their prime contractors, according to Dr. Walsh, which tends to mean a slow sales cycle. However, the systems are sometimes also used by small geophysical companies who contract to take the actual measurements and then provide the data to the end users. Sales to large entities are usually preceded by a rental evaluation period.

Dr. Walsh noted that while most sales are to large entities, Vista Clara does rent equipment to smaller companies, and in some cases has acted as the data collector itself, although it prefers to simply provide the equipment.

Marketing in this sector is highly specialized. Vista Clara attends 8-12 conferences annually, focused on interacting the hydrology scientists and their sponsors. The company also attends some conferences for vertical markets—for example, mining conferences in Vancouver and Australia. Vista Clara also publishes papers in peer-reviewed journals, as these articles are read by the customers the company is seeking, especially researchers and academics. Dr. Walsh observed however that publishing remained a challenge as company staff were usually fully committed with company projects.

Dr. Walsh said that the company faced three kinds of competitors:

- Existing established competitors. There is one primary established competitor, which is a state-owned French company with a product that is not cutting edge but which is supported by significant marketing help from the French government.
- Emerging competitors. There is one new company emerging in Australia.
- U.S. and European R&D groups that are trying to develop similar technology but have not yet successfully reached the market. These groups tend to be more focused on academic interests.

Vista Clara retains some key advantages, according to Dr. Walsh. The technology is hard to develop, although once developed it is easy to re-apply in different form factors. Dr. Walsh said that the company had used export services from the Commerce Department, with mixed results. The process had helped the company to acquire some customers in Denmark.

Patents and Other Intellectual Property

Vista Clara is the assignee for the U.S. patents listed in Table E-9.

Operations

Vista Clara generates income from its NMR hydrogeologic instruments, and exports are driving its sales success. Vista Clara reported recently that it has won four of its last five competitively bid proposals in China,

TABLE E-9 Vista Clara Patents

Patent Number	Patent	Year
8,816,684	Noise canceling in-situ NMR detection	2014
8,736,264	NMR logging apparatus	2014
8,581,587	SNMR pulse sequence phase cycling	2013
8,451,004	Multicoil low-field nuclear magnetic resonance detection and imaging apparatus and method	2013
RE43,264	Multicoil NMR data acquisition and processing methods	2012
7,986,143	Multicoil low-field nuclear magnetic resonance detection and imaging apparatus and method	2011
7,466,128	Multicoil NMR data acquisition and processing methods	2008
6,160,398	Adaptive reconstruction of phased array NMR imagery	2000

SOURCE: U.S. Patent and Trademark Office.

the most recent of which resulted in the sale of three GMR surface NMR instrumentation systems.[57]

SBIR/STTR and Vista Clara

Between 2003 and 2014, SBIR/STTR funded 14 projects with Vista Clara, Inc. amounting to nearly $5.5 million. Of this amount, DoE provided approximately 64 percent, DoD 21 percent, and NSF the remaining 14 percent. The company has received one Phase I and one Phase II STTR award from DoE.

Vista Clara's first SBIR award was a 2003 Phase I NSF award for adapting medical MRI technology for use in groundwater characterization. This was followed by other Phase I awards from DoD and then by a 2005 Phase II NSF award for $500,000 which transformed the company. It now no longer had to rely entirely on other companies for revenues, and could move forward to develop its first product.

By the end of the first Phase II award, the technology was good enough to collect data, and a customer in Germany was prepared to pay for a product in semi-finished format. Dr. Walsh said that he sold his house to raise the money to build the product.

Starting in 2008, Vista Clara received further Phase II SBIR and STTR awards from DoE, which have according to Dr. Walsh allowed it to gain

---

[57]"Vista Clara secures leading position in China," http://www.vista-clara.com/news/vista-clara-secures-leading-position-in-china/.

substantial ground on its competitors and develop fully finished products. Funding for the company's second product, the Javelin, came during this period.

Phase IIB funding at DoE was for a project to develop accustom cable for down-hole logging. Vista Clara had sought $300,000 from DoE and had invested $75,000 of the company's capital, and although the DoE program did not require matching funds, Dr. Walsh believed this investment helped the company win the award.

Dr. Walsh said that in his view it was important to ensure that the company had created a finished or close to finished product by the end of Phase II, otherwise it would need to find new funding or commit its own resources to fill the gap. The Javelin project fit this model, as a finished product had been completed by the end of the Phase IIB award. The product was now in use in Australia and by the U.S. Geological Service. Companies should also be aware that new technology took time to develop a sustainable market—early adopters could be relied on to purchase a few initial units, but that subsequent sales could take a considerable time.

DoE's interest in Vista Clara technology stems from the agency's need to manage groundwater contamination more effectively. Facilities are currently spending hundreds of millions of dollars on soil and groundwater remediation, and Vista Clara technology offers significant upgrades on existing approaches, according to Dr. Walsh.

However, despite the funding and interest expressed through SBIR awards sponsored by the office of subsurface biology, Vista Clara has as yet made no sales to DoE. Dr. Walsh observed that it appears there is no clear connection between the SBIR program and DoE needs elsewhere in the agency. Thus while there is a topic every year on subsurface characterization and remediation, there are no follow-on contracts for SBIR winners. Vista Clara has won three Phase II awards to develop the NMR technology that the company now sells, but which is not in use at DoE. Contracts for remediation are awarded through a large prime contractor and there appear to be no incentives for the use of small/SBIR companies. This remains the case even though Vista Clara has good contacts at the National Lab near the Hanford remediation site.

Dr. Walsh said that he strongly supported DoE's set aside of part of the STTR budget to pay for articles in peer review publications, which often charged significant amounts. DoE allows labor costs for preparing articles, presenting at conferences, and publication charges for print journals, although these costs do have to be included in the initial proposal budget. Other agencies should follow DoE's lead in this area.

DoE has also recently begun to allow patent application costs up to a set limit. This is a very welcome initiative, according to Dr. Walsh, as the costs otherwise come directly out of the company's profit. At DoE, these can be charged as direct costs.

Dr. Walsh said that he believed DoE reviews in some cases rely too heavily on academic reviewers. He found that proposals could be downgraded if they did not include an academic partner. And while he did not object to partnering with academic institutions on occasion, he said that in most cases Vista Clara could have done a better job without them. In only a few out of the 7-8 partnerships formed for SBIR/STTR did the university add real value.

XEMED, LLC[58]

Xemed LLC is a private company founded in 2004 by Dr. William Hersman, Professor of Physics at the University of New Hampshire. Xemed is headquartered in Durham, New Hampshire, and has grown to 10-15 employees over 11 years. The company has broad expertise and IP in the production of hyperpolarized noble gases, and its mission is to develop these inhaled diagnostic agents which are capable of changing the management of respiratory diseases.

Dr. Hersman said that he had not originally intended to start a company, and was still primarily an academic. In the early years of his professorship, he was conducting research as a nuclear physicist at particle accelerators, where his interest migrated toward the solution of technical problems rather than investigating fundamental questions. In the mid-1990s, his work on hyperpolarized gases increasingly shifted from nuclear physics into medical applications, where he identified important new processes first theoretically and then, with the aid of grant funding, experimentally.

During this period he had served on an NIH SBIR review panel. He saw first-hand the process of seeking funding for transitioning academic research to small business. Realizing that the next phase of his research with medical imaging would require a well-engineered system for producing hyperpolarized gases, he felt incentivized to turn to STTR and SBIR as the appropriate funding source. The fundamental science of hyperpolarizing gas was by then becoming well understood following his own academic work, and he believed there would be important opportunities for training students by investigating applications of this technology in a medical environment. However, additional development work would be needed to make his work sufficiently robust for a clinical setting. He was aware that his R01 grants at NIH were coming to an end and were unlikely to be renewed at the level required for robust refinement of the technology and engineering of a robust, automated system. This prediction proved to be true, but SBIR/STTR provided a new source of funding. He found that the funding agencies saw his work as a

---

[58]Primary sources for this case study are the interview with Dr. William Hersman, CEO and founder, August 27, 2015, and a review of the Xemed, LLC website (http://www.xemed.com) and related company documents.

good fit for SBIR/STTR, and were eager to transition applications-oriented scientists into the SBIR/STTR program, avoiding an interruption in his effort.

SBIR and STTR provided the funding that allowed Xemed to demonstrate its technologies and then to complete the development work that underpinned the release of its initial commercial products. Along the way, this work also led to academic awards. In 2011 Prof. Hersman's PhD student won the International Society of Magnetic Resonance in Medicine's W. S. Moore Young Investigator Award by demonstrating new lung imaging capabilities of MagniXene®. Two years later in 2013 a PhD student from the University of Virginia won the same Young Investigator award for his work with MagniXene®.

Xemed has encountered some regulatory challenges in bringing its products to market. The FDA has declared that polarized gas is a drug, so the commercial path and regulatory path are intertwined. The company has also been waiting for a patent held by another group to expire (which it will do in spring 2016). This patent covers technology that is used in Xemed's products, so full commercialization has been delayed as a result.

Xemed is addressing regulatory issues, and believes that FDA approval will come—the company submitted a New Drug Application (NDA) in August 2015, providing details of the Phase 1, Phase 2 and two confirmatory Phase 3 clinical trials that have been completed. The company is confident that the technology will prove out in trials—Dr. Hersman noted that the drug is extremely safe—almost in the category of "generally recognized as safe," as it is just an inert gas—and Xemed has been able to show tangible benefit from the technology.

Technology

Xemed's goal is to develop inhaled diagnostic agents that can improve the standard of care for respiratory disease, specifically chronic obstructive pulmonary disease (COPD), asthma, and lung cancer. The company is therefore working to establish hyperpolarized gas as a clinically validated, FDA approved, and easily-produced diagnostic agent for magnetic resonance imaging of the lung's functional microstructure.

Current techniques for evaluating lung function include spirometry (which is non-imaging), x-ray and computer tomography (which use ionizing radiation) and ventilation scans which use a radioactive marker (^{133}Xe or ^{99m}Tc). In contrast, xenon is a component of air that can be extracted and purified. Xenon-129 (^{129}Xe) is one of the naturally occurring non-radioactive isotopes of xenon. Having a natural abundance of 26 percent, ^{129}Xe can be enriched using isotope-separation centrifuges. By magnetizing samples of this isotopically-enriched gas, Xemed produces its inhaled contrast agent MagniXene®, that permits MRI scans to provide much enhanced resolution over the inhaled radioactive marker and also permits much more detailed diagnosis of a range of functional pulmonary parameters than CT or x-ray.

For clinical applications, Xemed is targeting two primary markets: COPD and asthma. In both cases, bronchoscope-based therapies—ventilation management for COPD and bronchial thermoplasty for asthma—may reduce suffering and frequency of hospitalizations. Both techniques could benefit from a granular understanding of lung function. Xemed's latest ongoing clinical trials are designed to determine whether the inhaled contrast agent allows specialists to use these new techniques effectively. Recently, lung cancer researchers have also hypothesized that hyperpolarized MRI can be used to refine management of stereotactic conformal radiotherapy for lung cancer.

Hyperpolarized Noble Gases

Most magnetic resonance imaging (MRI) systems use proton nuclear magnetic resonance. A strong magnetic field aligns the spins of the hydrogen atoms within a patient's body, and the patient is subsequently exposed to a radio frequency pulse that perturbs the aligned atoms and releases electromagnetic radiation that can be detected and analyzed to understand structures in the patient's body. Because gases are 1,000 times less dense than water, there is insufficient number of gas molecules at standard temperature and pressure to produce a strong enough electromagnetic signal from a gas using ordinary NMR techniques.[59]

Using spin exchange optical pumping (SEOP), however, it is possible to align the spins of noble gases by using polarized light to transfer angular momentum to the gas atoms. Circularly polarized infrared laser light excites electrons in an alkali metal vapor, such as cesium or rubidium. Collisions between metal electrons and noble gas nuclei ^{129}Xe transfer angular momentum to the gas. Nitrogen is used to prevent fluorescence of the polarized alkali metal which could de-polarize the gas. Using SEOP, hyperpolarized noble gases with 10^5 times the number of spin aligned ^{129}Xe gas nuclei than seen at standard temperature and pressure are possible.[60]

Hyperpolarized noble gases offer the potential for functional (as opposed to structural) imaging of lung tissue using magnetic resonance imaging (MRI). The challenge of producing sufficiently large amounts of hyperpolarized gas at sufficiently high levels of spin alignment, however, has limited the adoption of such gases as tools for managing lung disease. Also, because

---

[59]Jason Leawoods, *et al.* "Hyperpolarized ^3He Gas Production and MRI Imaging of the Lung," *Concepts in Magnetic Resonance*, vol. 13, no. 5 (2001), p. 277, http://onlinelibrary.wiley.com/doi/10.1002/cmr.1014/abstract.

[60]F. William Hersman, *et al.* "Large Production System for Hyperpolarized ^{129}Xe for Human Lung Imaging Studies" *Academic Radiology*, vol. 15, no. 6, (2008), pp. 683-684, http://www.ncbi.nlm.nih.gov/pubmed/18486005.

hyperpolarized noble gases (especially xenon) tend to de-polarize over time, imaging systems using such gases would require development of on-site production systems.

Xemed has developed self-contained robust and automated production systems for hyperpolarized noble gases that include a number of refinements to the standard SEOP approach. It has improved the polarization apparatus by developed techniques for narrowing the spectra of the laser sourcing the polarized light. The company has also improved the gas accumulation process by implementing robotic, low temperature systems to enhance gas trapping. Xemed has produced hyperpolarized ^{129}Xe with 64 percent of the sample spin aligned. However, because there is a tradeoff between output production rates (liter/minute) and spin alignment, Xemed's production system produces gas with spin alignment typically around 50 percent.[61]

Products

Hyperpolarized noble gases allow evaluation of a broad range of clinically important lung characteristics. At present, Xemed's collaborators at academic hospitals in the United States and Canada are developing protocols for their use in estimating lung ventilation, alveolar size, small airway dimension, exchange with red blood cells, and other parameters of lung function.

Xemed is developing two products, MagniXene® and MagniLium, and their associated production systems. Over the past fifteen years, the bulk of research in clinical applications of hyperpolarized noble gases has focused on ^{3}He. Xemed is also developing a ^{3}He-based product called MagniLium, but because ^{3}He is an artificial stable isotope whose only source comes from maintaining the nuclear weapons stockpile, Xemed management expects that ^{129}Xe will, in the long term, be the gas adopted as the inhaled diagnostic agent of choice, because it is naturally available in air.

Hyperpolarized Xenon—MagniXene®

MagniXene® is hyperpolarized ^{129}Xe gas. As a diagnostic drug, it enables quantification of pulmonary function within different regions of the lungs. Xemed envisions two applications for MagniXene®. As a tool in clinical care, MagniXene® could improve performance of procedures such as bronchial thermoplasty for severe asthma or ventilation management for advanced COPD. Also as a drug development tool, pharmaceutical companies could learn more

---

[61]F. William Herman, "Xemed LLC is developing magnetized gas," http://www.slideshare.net/changezkn/xemed-llc-is-developing-magnetized-gas-hyperpolarized-xenon.

about the effectiveness of their therapeutic pulmonary drugs by obtaining more information from the lungs of clinical trial participants, potentially accelerating FDA approval of their drugs and reducing time to market.

To do this, Xemed has focused its development efforts on a robust, automated, compact, self-contained MagniXene® production system that it calls the XeBox. Now in its sixth generation of technical refinement, the Xebox embodies five patents licensed from the University of New Hampshire. Beginning with a gas bottle filled with isotopically enriched ^{129}Xe, this system mixes the gas with rubidium vapor and other gases, illuminates it with laser light performing the hyperpolarization, separates and accumulates the product cryogenically, and loads the resulting gas into a breathing bag for use by a patient. This takes about 20 minutes. Xemed has also developed related equipment prototypes for use with the gas such as a chest coil with 32 receiving elements mated to an asymmetric birdcage transmit coil.

Xemed has completed two clinical trials of MagniXene®. The Phase 1 trial partnered with researchers at the Brigham and Women's Hospital and Harvard University to evaluate using ^{129}Xe as a contrast agent to study patient safety and preliminary indications of efficacy. In the Phase 2 and Phase 3 trials, Xemed partnered with researchers at the University of Virginia to assess regional lung function in patients suffering from COPD and asthma. Xemed is now recruiting for a Phase 2 study of therapeutic efficacy with Washington University in St Louis to evaluate using ^{129}Xe to guide bronchial thermoplasty for severe asthmatics.[62]

Hyperpolarized Helium—Magnilium

The production system for MagniLium, Xemed's ^{3}He hyperpolarized noble gas product, is the HeliBox-Z100, designed for larger scale production runs with an 8.5 liter monolithic aluminosilicate cell in a pressure and temperature controlled environment. The second generation Helibox achieved helium quantities as large as 50 liters per day, and the third generation confirmed that polarization levels can reach as high as 60 percent. Based on the company's patented spectral narrowing laser technology, the fourth generation

---

[62]"Quantifying Collateral Perfusion in Cerebrovascular Disease-Moyamoya Disease and Stroke Patients," ClinicalTrials.gov, 2013, https://www.clinicaltrials.gov/ct2/show/NCT01419275?term=Xemed&rank=2; "MagniXene MRI Use in Patients With Asthma and COPD to Assess Regional Lung Function by Delineating Ventilation Defects (HXe-VENT)," ClincialTrials.gov, 2015, https://www.clinicaltrials.gov/ct2/show/NCT01833390?term=Xemed&rank=3; "Bronchial Thermoplasty for Severe Asthmatics Guided by HXe MRI (HXe-BT)," ClinicalTrial.gov, (Incomplete) https://www.clinicaltrials.gov/ct2/show/NCT01832363?term=Xemed&rank=1.

system will probably show the highest level of ^3He polarization yet reported. The company has not yet undertaken clinical trials for MagniLium.

Patents and Other Intellectual Property

Xemed, LLC does not have any U.S. patents assigned to it. However, F. William Hersman, the founder and CEO of Xemed, LLC, is the inventor of the patents (listed in Table E-10) related to production and use of polarized noble gases. With the exception of U.S. Patent 7,719,268, these patents are assigned to the University of New Hampshire.

Internationally, one European patent has been awarded with another pending both in Europe and elsewhere.

Collaboration

Xemed maintains research relationships with various university laboratories and research hospitals. Xemed worked with the University of New Hampshire and Mass General Hospital in the development of hyperpolarized noble gas production systems and chest coil development. In developing imaging protocols for use with its MagniXene® product, Xemed has performed or is performing clinical trials with the University of Virginia Health System, Washington University in St. Louis School of Medicine, and two imaging centers in Ontario Canada, the Robarts Research Institute and the Thunder Bay Regional Research Institute. A pharmaceutical company funded a pilot study with Xemed to use its Xenon-based diagnostic to assess its application in evaluating the efficacy of a pulmonary drug recently under development.

Business Model

Xemed, LLC has received support from SBIR/STTR funding, other grants, and revenue from operations (see SBIR/STTR section). Xemed has also generated revenue from the sale of professional services. The company reports

TABLE E-10 Xemed, LLC Patents

Patent Number	Patent	Year
8,405,022	Thermal management technology for polarizing xenon	2013
7,928,359	Thermal management technology for polarizing Xenon	2011
7,769,068	Spectral-narrowing diode laser array system	2010
7,719,268	Apparatus and method for polarizing polarizable nuclear species	2010
7,281,393	Method and apparatus for accumulating hyperpolarized xenon	2007
6,949,169	Apparatus and method for polarizing polarizable nuclear species	2005

SOURCE: U.S. Patent and Trademark Office.

on its web site that it had raised $7 million in cumulative non-dilutive capital through competitive research proposals since its founding.[63] No date is given.

Xemed has received non-monetary commercialization guidance from various sources. For example, in 2010-11, the company participated in the National Institutes of Health Commercialization Assistance Program (CAP), a 9 month mentoring program designed to help participants understand their commercialization options and develop a market- and customer- driven commercialization plan.[64] In 2012 Xemed participated in the Niche Assessment Program.

Xemed has also received grants from non-federal sources. For example, in 2006, it received funding from the New Hampshire Innovation Research Center.[65]

Xemed is now beginning to gain traction in the marketplace, a process that Dr. Hersman believes will accelerate sharply in 2016 when lingering patent issues are no longer relevant.

The company is already finding new interest from researchers wanting to buy machines to make polarized gas in support of their NIH-funded projects. Interest is also growing from universities seeking to buy machines to polarize gases. However, these tools still cost on the order of $600,000-$1 million per machine, as they are still hand-assembled by PhDs. Dr. Hersman believes that this could be reduced by 50 percent with a higher volume of sales.

Drug companies also are interested in using Xemed technologies. Vertex pharmaceuticals recently conducted a cystic fibrosis study using Xemed's tools at the University of Virginia, where Novartis and Teva are also conducting studies using an in-house polarizer of ^{3}He. There are in addition two drug studies under way in the United Kingdom with in-house built polarizers.

Because Xemed's machines can potentially impact a wide range of diseases and interventions, the company believes that it will gain rapid traction in the marketplace as individual treatments are adopted that use the technology. For example, bronchial thermoplasty is a recently developed approach that works by microwaving smooth muscles in airways. Xemed technology is used to guide the intervention, and as the technology becomes more widely available, Xemed can expect to find further demand for its products.

Similarly, Dr. Hersman believes that Xemed technologies can be used to provide images and biomarker data that are uniquely sensitive, to show very

---

[63]Xemed, "Corporate Achievements," http://www.xemed.com/company/achievements/.

[64]"Xemed selected to participate in the 2010-2011 NIH SBIR Commercialization Assistance Program," October 1, 2010, https://www.xemed.com/2010/10/xemed-selected-to-participate-in-the-2010-2011-nih-sbir-commercialization-assistance-program/; "Commercialization Accelerator Program (CAP)," August 14, 2015, https://sbir.nih.gov/cap.

[65]"NHIRC Highlights," http://www.nhirc.unh.edu/pdf/2015-NHIRC%20Impact%20Report.pdf, p. 8.

specific regional characteristics of lung disease. Confirming this view will require further tests and trials, as yet only a limited number of doctors have seen the Xemed images and developed ways to adjust the treatment regimen as a result. Benefits have therefore not yet been quantified, and the value to health care funders such as insurance companies is not yet well established.

SBIR/STTR

Between 1990 and 2013, SBIR/STTR funded 27 projects with Xemed, LLC amounting to nearly $15.6 million in R&D support. NIH provided 86 percent of Xemed's SBIR/STTR funding, and DoE provided the remaining 14 percent. Twenty-one percent of funding was STTR, which accounted for 13 of the 27 funded awards.

STTR was the initial funding for the company. Xemed had applied for three STTR awards to improve three different aspects of the apparatus. In fact, Xemed was founded only when STTR funding was awarded. The company received two Phase I awards and then a subsequent Phase I the next year. All three transitioned to Phase II.

STTR and SBIR funding allowed the company to make important technical breakthroughs, leading eventually to an improvement over the prior technology by a factor of 100, according to Dr. Hersman. STTR funding has also been used to demonstrate medical imaging aspects of the gas with a wide range of academic partners. Xemed has had at least twenty different research collaborators with whom they have worked to improve various aspects of the technology, and to develop and prove out specific applications.

However, despite a number of successes Xemed had also had negative experiences with STTR. In 2007-2008 the company had received a DoE STTR award to re-examine the utility of hyperpolarized ^{3}He for nuclear physics. A contracting officer within DoE believed that STTR awards could not have an academic PI. DoE as a result halted the grant, which almost destroyed the company. Dr. Hersman had to mortgage his house to fund the company. And while the DoE SBIR/STTR staff worked to address the problem, it appeared that DoE simply had no mechanisms available for resolving the problem. In the end, the project was converted to an SBIR award, and a different PI was assigned from within the company. The delay caused milestones to be missed, but DoE has continued to support the research and Xemed work in this area is currently funded through a Phase IIB award from DoE.

On the other hand, the initial STTR partnership with the University of Virginia went very smoothly, according to Dr. Hersman, despite some initial ambiguity about who could be PI—the need to demonstrate a "special relationship" with the company was addressed by giving the PI (at the university) an appointment on Xemed's scientific advisory board. Since this initial STTR, all medical research with the University of Virginia channeled through Xemed has been proposed through the SBIR mechanism.

Xemed has participated twice in the Commercialization Accelerator Program (CAP) run by Larta, and has also received a market assessment from Foresight. Overall Dr. Hersman said that these programs were helpful in creating a strategy for regulatory advancement and a more effective approach to commercialization.

Xemed works effectively across SBIR/STTR agencies by developing new applications for its technology: DoE is interested in the technologies developed initially for medical applications, with a view to utilizing them to enhance efficiency within DoE's neutron scattering and nuclear physics projects. The company has also branched out into defense applications, and two recent submissions crafted by Xemed to Broad Agency Announcements from DoD led to funding. Due to the preference of the Sponsor, they became University-led projects.

Recommendations for Improvement

Overall Dr. Hersman believed that recent developments such as the emergence of Phase IIB awards have been positive. He also continues to be impressed by the NIH review process.

Dr. Hersman also recommended that NIH improve the connections between Institutes. For example, a Xemed NIH Phase I award was initially assigned to NHLBI but was then funded by NCI. The latter then did not wish to fund Phase II. At that point NHLBI would not pick up the project, even though its score was well within the NHLBI's Phase II payline. That particular project was never funded resulting in hardship especially for the academic partner.

XIA, LLC[66]

XIA LLC (originally X-Ray Instrumentation Associates) is a private company founded in 1988 by William Warburton. The company invents, develops and markets advanced digital spectrometers for x-ray, gamma-ray, and other radiation detector applications in university research, national laboratories and industry. XIA is headquartered in Hayward, CA, and generates income from the design, development and marketing of spectrometers.

XIA was founded by Dr. Warburton as a sole proprietorship in 1988, following a career as a materials researcher, including a period employed at the Stanford Synchrotron Research Laboratory (SSRL) where he was a beamline

[66]Primary sources for this case study are the interview with Dr. William Warburton, CEO and founder, August 24, 2015, and a review of the XIA web site (http://www.xia.com) and related company documents.

scientist. He left when SSRL shut down for a year to make needed repairs, and founded XIA. The company emerged in earnest when Dr. Warburton's first Phase I SBIR award from NIH in 1991 was followed by Phase II and he hired employees to assist with the research.

The company became sustainable after the SBIR-funded development of electronics to control spectrometers, replacing the difficult to tune and expensive to maintain analog controls that had previously been industry standard.

XIA has also responded to DoE SBIR topics that call for tools related primarily to x-ray and nuclear electronics, according to Dr. Warburton. This approach worked moderately well for a period, providing sufficient revenue to support core company R&D operations. The resultant instruments generated sales to national and international labs, primarily of digital spectrometers for both synchrotron x-ray spectroscopy and for medium sized nuclear experiments. A typical product generated perhaps $200,000 annually in revenues for between 5 years and 10 years.

Until recently, the company depended on SBIR or Broad Agency Announcement (BAA) funding to support its advanced R&D activities, using income derived from sales to support new product development. The company currently derives about 75 percent of its income from product sales, with the rest coming from SBIR and BAA grants and from commercial contracts.[67]

The company maintains research relationships with a broad range of academic, government, and corporate entities such as UC Davis, University of Texas at Austin, Michigan State University, Pacific Northwest National Laboratory, Los Alamos National Laboratory, Lawrence Livermore National Laboratory, Institute for Nuclear Physics (Germany), Radiation Protection Bureau, Health Canada, Alameda Applied Science Corporation, and IBM, to name only a few.

XIA Technologies

Radiation Data Detector Acquisition Systems

XIA develops digital data acquisition and processing systems for x-ray, gamma-ray, and other radiation detectors. The company's core technology combines digital signal processors (DSP) with field programmable gate arrays (FPGA) and—in various forms—has enabled XIA's portfolio of high speed spectrometers. The FPGA performs and manages data acquisition and storage (i.e. pulse detection, filtering, pileup inspection and coincidence inspection) and

[67]XIA LLC, https://www.linkedin.com/company/xia-llc.

the DSP performs higher level post processing analysis (i.e., baseline correction and pulse shape analysis). The FPGA stores input signals to different parts of the system memory based on external interrupts generated by the sensors.

XIA has applied this architecture to a range of problems, in both industry and basic research. For example, XIA x-ray spectrometers have been used in metal sorting facilities: exposed to x-rays, different metals fluoresce in different parts of the spectrum, and XIA tools can identify which metals are fluorescing. DXP systems are then used to analyze the data from x-ray detectors and guide mechanical systems to sort the different types of scrap metal.

A nuclear application example is in low background gamma spectroscopy. In health physics, nuclear waste management, and nuclear materials and weapons security, the ability to detect small amounts of gamma radiation against background noise is vital. A XIA PXI-based processor can be used to veto signals that fail pulse shape or coincidence tests and so remove unwanted background events.

Other applications include handheld metal detectors using x-ray fluorescence, high-rate gamma spectroscopy for assaying spent nuclear fuel, discrimination of alpha, beta, gamma, and neutron radioactivity for detectors sensitive to the full range of radiation events, and synchrotron-based spectroscopy for characterizing materials properties in pharmaceutical, engineering, and material science.

Product Architectures

XIA's product line falls into three main digital data acquisition architectures: DXP (Digital X-ray Processor), DGF (Digital Gamma-ray Family), and Ultra-Lo (ultra-low background alpha particle detectors). They allow researchers to store, count, and analyze (height, shape, etc.) the analog signals captured by various different sorts of radiation sensors.

The full line of XIA products includes 13 different products. All can be further customized to particular customer needs. Depending on the system characteristics, XIA's data acquisitions systems range in price from $750 to $60,000.[68]

DXP

The DXP family of products implements XIA's core FPGA—DSP innovation. A field programmable gate array (FPGA) provides the front end shaping of the input signal steps generated by the sensor array and extracts their

---

[68]XIA, LLC, https://www.linkedin.com/company/xia-llc.

amplitudes in real time, while a digital signal processor provides corrections to improve energy resolution and stores the resultant values in a spectrum. Because the processing dead time per signal step in DXP processors is essentially zero, extremely high count rate (up to 1 million counts per second) are possible. The DXP architecture is available in products ranging from low cost OEM cards for handheld and bench top applications to PXI-based standalone modules for ultra high rate counting in, for example, synchrotrons or industrial control applications.

DGF

The DGF architecture extends the DXP architecture. With a FIFO memory for digital signal capture and a flexible, two-level triggering system that can span multiple modules, the DGF's digital signal processor—in addition to the pulse height measurement performed by DXP systems—can also perform real time analysis of pulse shape. For example, incoming data can be processed and sorted according to pulse shape characteristics such a risetime or falltime. The DGF product line provides solutions to a wide range of extremely demanding pulse processing applications in the areas of nuclear physics, strip detectors, and very high resolution gamma-ray spectroscopy.

ULTRA-LO 1800

The Ultra-Lo 1800 is based on the DGF architecture and designed to measure the alpha particle emissivity of solid materials. Using dual channel pulse shape analysis, the Ultra-Lo 1800 is able to distinguish between alpha particles emitted by the sample under test and alpha particles generated elsewhere in the instrument. Rejecting the latter, the Ultra Lo 1800 can detect background rates as low as 0.0001 alpha particles/cm2 per hour. This is a factor of 50 or more time lower than can be achieved using the current state of the art proportional counting systems. The Ultra Lo 1800 was developed to improve quality control processes in the semiconductor manufacturing industry with SBIR funding from NIST and DoE.[69]

Patents and Other Intellectual Property

XIA is not the assignee of any U.S. patents. However, the patents (listed ion Table E-11) assigned to William Warburton, the CEO of XIA, are

---

[69]SBIR Success Story: XIA, LLC, http://www.nist.gov/tpo/sbir/sbir-success-story-xia.cfm.

TABLE E-11 Patents Assigned to William Warburton, CEO of XIA

Patent Number	Patent	Year
7,966,155	Method and apparatus for improving detection limits in x-ray and nuclear spectroscopy systems	2011
7,342,231	Detection of coincident radiations in a single transducer by pulse shape analysis	2008
7,065,473	Method and apparatus for improving resolution in spectrometers processing output steps from non-ideal signal sources	2006
6,732,059	Ultra-low background gas-filled alpha counter	2004
6,609,075	Method and apparatus for baseline correction in x-ray and nuclear spectroscopy systems	2003
6,590,957	Method and apparatus for producing spectra corrected for deadtime losses in spectroscopy systems operating under variable input rate conditions	2003
6,587,814	Method and apparatus for improving resolution in spectrometers processing output steps from non-ideal signal sources	2003
6,169,287	X-ray detector method and apparatus for obtaining spatial, energy, and/or timing information using signals from neighboring electrodes in an electrode	2001
6,125,165	Technique for attenuating x-rays with very low spectral distortion	2000
5,873,054	Method and apparatus for combinatorial logic signal processor in a digitally based high speed x-ray spectrometer	1999
5,870,051	Method and apparatus for analog signal conditioner for high speed, digital x-ray spectrometer	1999
5,774,522	Method and apparatus for digitally based high speed x-ray spectrometer for direct coupled use with continuous discharge preamplifiers	1998
5,684,850	Method and apparatus for digitally based high speed x-ray spectrometer	1997
5,646,488	Differential pumping stage with line of sight pumping mechanism	1997

SOURCE: U.S. Patent and Trademark Office.

solely licensed to XIA and potentially applicable to any hardware or software developed by XIA.

SBIR/STTR

Between 1990 and 2013, SBIR funded 53 projects with XIA LLC amounting to nearly $14.3 million. DoE provided approximately 76 percent, NIH 21 percent, and the Department of Transportation the remaining 3 percent.

Annual funding was close to $1 million from SBIR/STTR between 2007 and 2012. It has since declined significantly.

In general, Dr. Warburton said that SBIR/STTR had been critical to the foundation and growth of the company. These funds would not have been available from other sources.

However, Dr. Warburton had now come to believe that simply responding to available topics was not always in the company's best long term interest. The company's original business model had led to commercialization at approximately the level of agency SBIR investment, and so produced a steady-state business. But this ignored the opportunity cost to XIA of time spent simply maintaining the company instead of pursuing opportunities for greater growth.

While there are risks involved in taking a different approach, Dr. Warburton believes that the benefits can be considerably greater. He noted that, while a prototype of XIA's Ultra-Lo product emerged successfully following two small SBIR awards (DoE Phase I and NIST Phase I and Phase II), The company then invested approximately $3.5 million in the product over a period of ten years, to develop instruments with a much larger potential market selling for approximately $80,000 each. Market research suggested that XIA would sell 50 instruments a year, and he believes that the company will eventually reach that goal though perhaps not for some years. The company is currently waiting for NIST to produce a standard which will open the door to the marketplace. Until then, less sensitive existing instruments can be used and hence to do not need to be replaced.

Metrics. Dr. Warburton also observed that using commercialization as the only metric for assessing the success of SBIR award was misguided. XIA has sold maybe $10 million to $20 million in instruments for synchrotrons. The latter cost $500 million each to build and perhaps $200 million annually in running costs, but a large percentage of the research undertaken with these systems required instruments such as XIA's. Synchrotron x-ray fluorescence experiments would not run at all without them, and overall productivity (and hence return on investment) would be a fraction of what it was today. Similarly, XIA develops instruments for measuring background radiation that have been used for validating compliance with nuclear testy-ban treaties—another market with minimal sales but large social impacts.

Topics. XIA is seeing fewer topics that are potentially viable under current SBIR evaluation procedures, according to Dr. Warburton. While DoE scientists continue to seek tools and instruments to assist in their research, these generally have extremely limited commercial potential and hence fail DoE's "return on investment" (as measured only by instrument sales) criteria. For example, one recent topic was clearly designed to develop an instrument for use within the four accelerators that exist worldwide. This has almost no commercial potential.

Dr. Warburton said that, in the main, DoE topic managers still appeared to view SBIR/STTR as a tax on their research funding, and so wish to use it to provide tools or technologies that could be used to further their own scientific interests and programs. They have no interest in commercial potential, and he saw no evidence that topics were reviewed for commercial potential before being published. More generally, it did not appear that topics were subject to significant screening or review.

Many DoE topics are highly specific, tuned to the specific technical needs of topic managers. The agency has now started adding broader topics and does occasionally fund them. XIA did win a Phase I for a broader topic, although it did not go to Phase II.

Commercialization review. Dr. Warburton sees a substantial disconnect between the demands of topic managers focused exclusively on science and their technical needs and commercialization review. He found difficult to pass both reviews. His personal view was that small instrument sales that supported the national laboratories' missions were in the national interest and that this class of SBIR topic should be given evaluation criteria that appropriately reflect their values to those missions. Or, if the DoE only wants responses capable of large commercial returns, it should revamp its calls for proposals to bring them into conformance.

DoE now appears to require projections of sales quite far downstream. These future expected sales have to be large enough to recover the current SBIR investment plus provide an annual internal rate of return of 8 percent. This is a substantial hurdle, especially for products which are high risk and where markets are small—it was not clear to Dr. Warburton that any company providing high tech, low volume scientific instruments would ever meet this hurdle rate. He also wondered whether DoE has ever compared actual commercial outcomes in funded Phase II projects to the outcomes projected in the submitted commercialization plans in order to evaluate whether the present methodology actually has any predictive capability or is just an exercise in creative writing.

Review process. More generally, Dr. Warburton said that he had been an NIH SBIR reviewer and saw a number of features of the NIH process that might be beneficially adopted at other agencies. In particular, he believed that the face to face (or phone conference) meeting of the review panel provided a strong boost to the effectiveness of the review overall. In particular, the discussions between the reviewers quickly exposed the strengths and weaknesses of the arguments of both proposers and reviewers. At DoE the reviewers never connect, and as result reviewers can misunderstand the proposal—in both positive and negative ways—without having to justify their criticisms to their peers on the panel. In one particularly glaring case, XIA experienced a reviewer who was clearly commenting (negatively) on a non-XIA proposal.

Dr. Warburton also noted that there was no appeal process at DoE, and no possibility for resubmission (as at the NIH). He was therefore a strong proponent of the idea that companies be given an opportunity to respond (briefly—1 to 2 pages maximum) to reviewer comments before final decisions were made.

Operations. Dr. Warburton noted that the DoE payment system is excellent.

STTR

XIA has not had good experiences with the STTR program, Dr. Warburton said. For example, a collaboration with Brookhaven National Laboratory worked out poorly, with no accountability for the project at the lab. The project was developed to help measure carbon levels in the soil, focused on evaluating farming processes that could potentially remove carbon from the atmosphere. The Lab's main role was to develop a vehicle for safely moving the instrument, which included a neutron generator) across a field to be measured, but did not meet project objectives nor produce the vehicle within the time frame of the project.

National Labs have few incentives to cooperate fully with small businesses, Dr. Warburton observed. In the best of cases, the lab scientists involved saw STTR as a means of supporting their own research program, in exchange for providing the company with technical support. In other cases, though, lab staff saw the program simply as a means to generate funds and had no interest in commercial outcomes or even their partner's interests

An ongoing collaboration with Lawrence Livermore National Lab (within the context of an SBIR grant) is proving more successful. It linked to a scientist whose life's work is aimed at moving his technology out into the world. He provided access to detectors and sources and lots of feedback. In exchange, XIA supplied him with next generation electronics for his experiments. The collaboration had now lasted ten years, advanced the state of the art, and should be seen as quite successful.

XIA has not worked collaboratively with the national labs outside the SBIR/STTR program. It does provide customized instruments to lab staff, but on a contract basis. Sometimes this results in joint scientific publications. Dr. Warburton noted that each national lab had its own culture(s); XIA has worked quite successfully, for example, with Pacific Northwest National Lab generally, with a few departments at Lawrence Livermore National Lab, but essentially not at all with Lawrence Berkeley National Lab, even though it is the closest of the three.

Appendix F

Annex to Chapter 3: Agency-level Data

The following section uses data provided by the awarding agencies to describe patterns for awards and applications. Because agencies did not provide identical data, the section does not provide the same analysis for each agency.

DEPARTMENT OF DEFENSE

Phase I STTR

The overall numbers of DoD Phase I STTR applications grew steadily from 2003 to 2006 but have declined since 2008 (see Figure F-1).

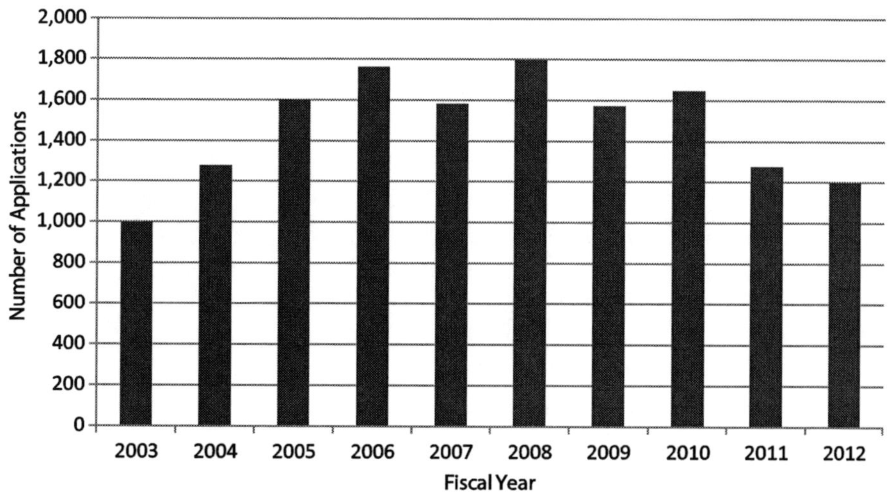

FIGURE F-1 Phase I STTR applications at DoD, FY2003-2012.
SOURCE: Data provided by DoD.

The decline in STTR applications after 2008 parallels a decline in SBIR applications, but it is steeper. The number of applications declined by one-third, from a peak of 1,796 in 2008 to a low of 1,198 in 2012, while the number of SBIR applications declined by about 15 percent.[1] Applications and awards did not track closely at DoD. In particular, the number of awards did not mirror the high number of applications received during FY2005-2010 (see Figure F-2). As a result, the success rates for DoD STTR applications varied substantially from a high in FY2004 to a low in FY2008, but then largely leveled out again from FY2009 through 2012 (see Figure F-3). Success rates for STTR applications during the study period averaged 21.5 percent. In comparison, the average success rate for DoD's SBIR program was somewhat lower, at 14.2 percent.[2]

Phase II STTR

DoD was particularly affected by the changes mandated under STTR program reauthorization. The requirement that all Phase I winners be permitted to apply for Phase II forced DoD to change its selection procedure substantially

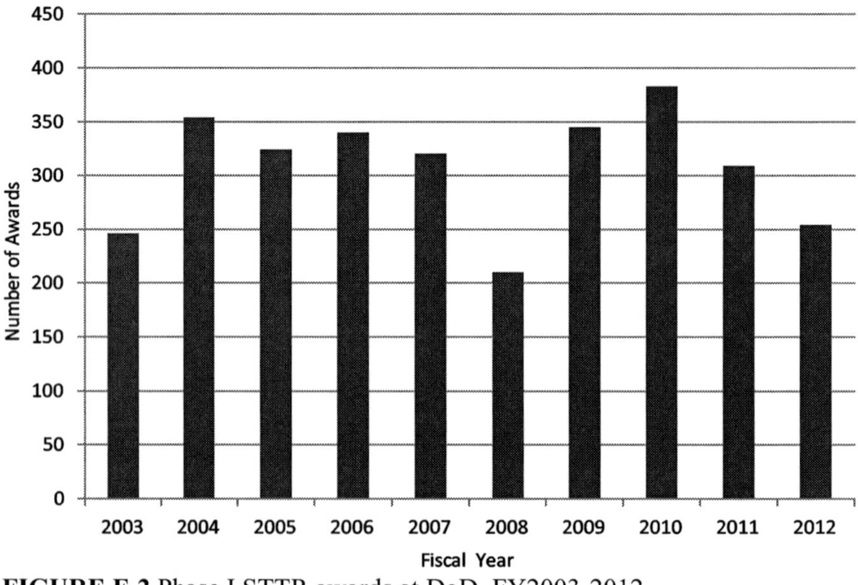

FIGURE F-2 Phase I STTR awards at DoD, FY2003-2012.
SOURCE: Data provided by DoD.

[1]National Research Council, *SBIR at the Department of Defense*, Washington, DC: The National Academies Press, 2014, p. 28. Effective July 1, 2015, the institution is called the National Academies of Sciences, Engineering, and Medicine. References in this report to the National Research Council or NRC are used in an historic context identifying programs prior to July 1.
[2]National Research Council, *SBIR at the Department of Defense*, p. 30.

and led to a considerable increase in Phase II applications in FY2011 and FY2012 (see Figure F-4).

The increase in Phase II applications did not translate into an increase in the number of awards, which, in fact, declined quite sharply during the same time period (see Figure F-5).

Phase II success rates changed after the 2011 reauthorization (see Figure F-6). Previously, DoD components selected which Phase I projects to invite to apply for Phase II and typically selected less than two per topic. These actions limited the number of Phase II applicants. After reauthorization, all agencies are required to permit all Phase I winners to apply for Phase II. According to Dusty Lang, the Navy STTR manager, this change both reduced success rates and added considerable administrative burdens to both the agency and the companies.[3] Reasons for the extremely high Phase II success rate in 2010 have not been determined.

STTR Awards by Component

Tables F-1 and F-2 show trends in STTR Phase I and Phase II awards by DoD component, respectively. The substantial decline in STTR Phase I awards at DoD after 2010 is driven largely by the sharp declines at Navy and Air Force and to a lesser extent at Army as well.

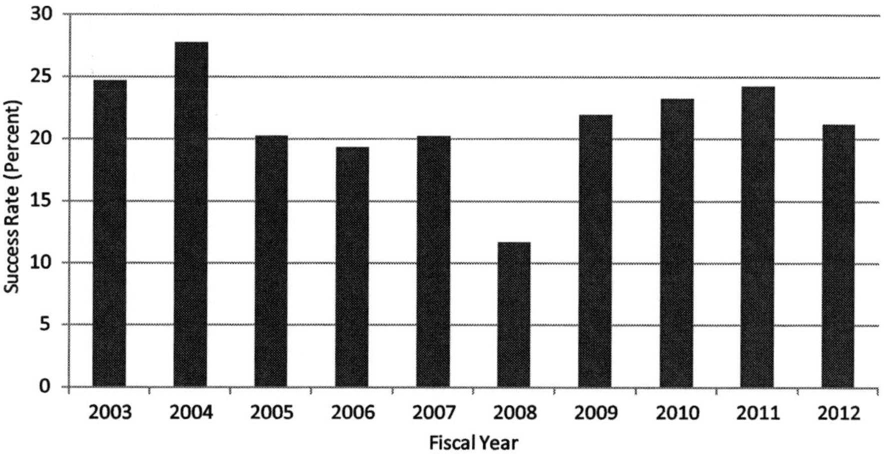

FIGURE F-3 Phase I STTR success rates at DoD, FY2003-2012.
NOTE: The success rate is the number of awards as a percentage of the number of applications.
SOURCE: Data provided by DoD.

[3]Interview with Dusty Lang, Navy STTR program manager, September 14, 2015.

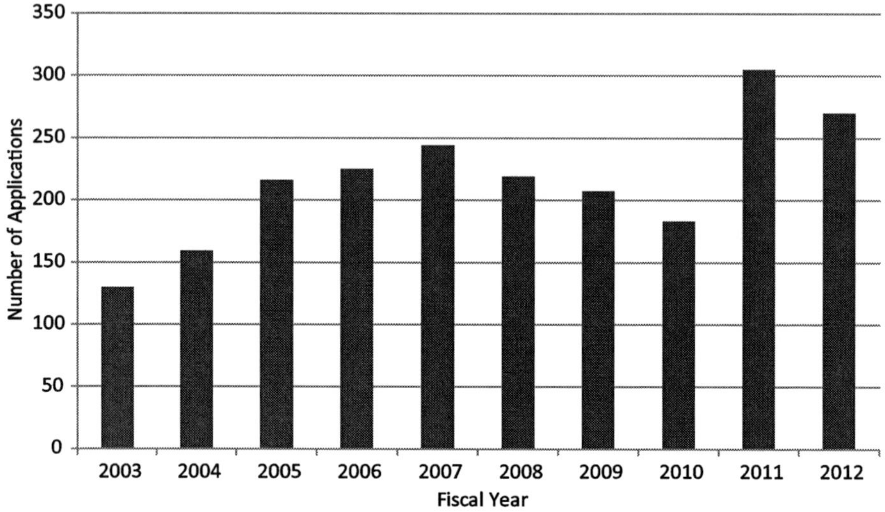

FIGURE F-4 Phase II STTR applications at DoD, FY2003-2012.
SOURCE: Data provided by DoD.

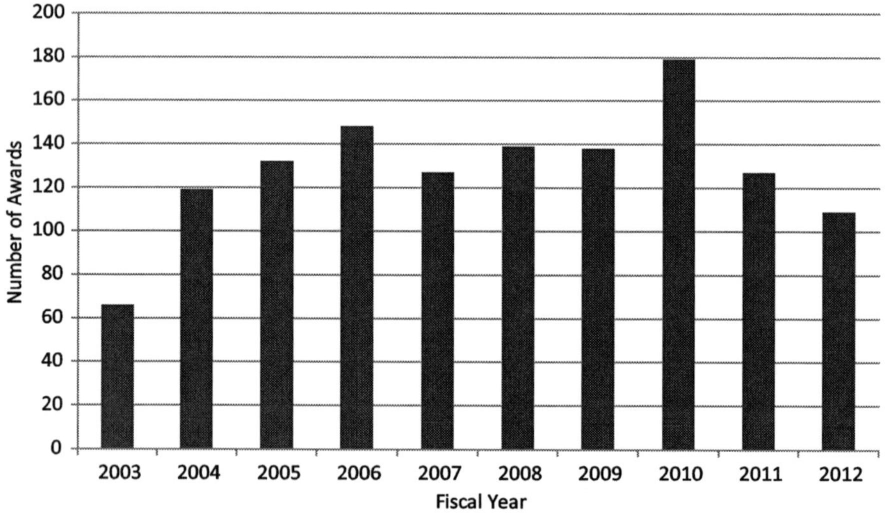

FIGURE F-5 Phase II STTR awards at DoD, FY2003-2012.
SOURCE: Data provided by DoD.

NATIONAL INSTITUTES OF HEALTH

Phase I STTR

NIH offers three deadlines annually for STTR awards. Unlike DoD, NASA, and DoE, it will consider funding any project that aligns with its

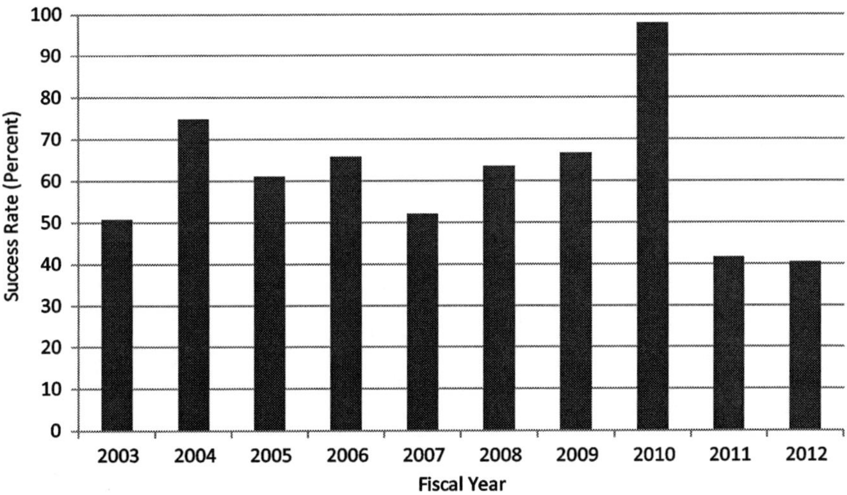

FIGURE F-6 STTR Phase II success rates at DoD, FY2003-2012.
SOURCE: Data provided by DoD.

TABLE F-1 Number of STTR Phase I Awards

	Fiscal Year										
Component	2005	2006	2007	2008	2009	2010	2011	2012	2013	2014	Total
Navy	96	116	95	93	120	151	117	62	39	59	948
Air Force	106	87	107	31	145	125	87	100	45	53	886
Army	48	79	41	63	6	64	63	42	35	36	477
MDA	30	28	24		25	25	19	12	29	24	216
DARPA	24	17	16	19	41	9		22	24	10	182
OSD	20	13	37	4	8	9	23	16	19		149
Component not recorded									34		34
Total	324	340	320	210	345	383	309	254	225	182	2,892

NOTE: The very small number of awards at several components in 2013 is not explained in the DoD annual STTR report, although it seems likely that it reflects tracking issues, as the number of award not assigned to any component was much higher in that year than in any other.
SOURCE: Data provided by SBA.

TABLE F-2 Number of STTR Phase II Awards

	Fiscal Year										
Component	2005	2006	2007	2008	2009	2010	2011	2012	2013	2014	Total
Air Force	49	43	38	29	46	66	28	36	5	43	383
Army	26	33	31	37	22	33	31	18	3	23	257
DARPA	8	6	7	10	8	15	3	2	3	16	78
MDA	16	16	12	9	8	13	6	12	6	12	110
Navy	33	42	33	52	43	41	44	13	5	22	328
OSD		8	5	2	8	4	3	5	1	4	40
SOE										1	1
Component not recorded			1		3	7	12	23	103	1	150
Total	132	148	127	139	138	179	127	109	126	122	1,347

NOTE: SBA data for 2013 appear to be misallocated, with almost all awards in the SBA database marked as blank for component.
SOURCE: Data provided by SBA.

mission, that is, the project does not have to address a specific topic from the solicitation. The number of Phase I applications to NIH declined between FY2007 and 2011, before expanding again in 2012-2014, and especially in 2014 (see Figure F-7). NIH averaged 647 Phase I STTR applications annually over the study period.

The total number of Phase I STTR awards at NIH fell during the first part of the study period, tracking a similar decline in SBIR awards. The number of awards began to increase again in 2012, especially in 2014 (see Figure F-8). Across the study period, NIH made an average of 120 awards annually.

Phase I STTR success rates at NIH remained relatively stable across the study period, varying from slightly less than 15 percent to more than 20 percent, and averaging 18.7 percent. This was close to the success rate for Phase I SBIR at NIH (see Figure F-9).[4]

Phase II STTR

Phase II applications at NIH are largely but not entirely driven by the number of Phase I awards made in the previous year. After rising sharply in 2005 and 2006 to a peak of 128 in 2007, the number of Phase II applications declined to about 70 in 2012 and 2013 before increasing again in 2014 (see Figure F-10). Given the substantial increase in Phase I awards made in 2014, it seems likely that the 2015 data will show another sharp increase in the number of Phase II applications.

[4]NIH RePorter database, Table 216.

Award numbers are more volatile than application numbers. Awards increased early in the study period, peaking at 46 in 2008 before declining to below 20 in 2013 and increasing again in 2014 to 37 (see Figure F-11).

Phase II STTR success rates were more volatile than those for Phase I,

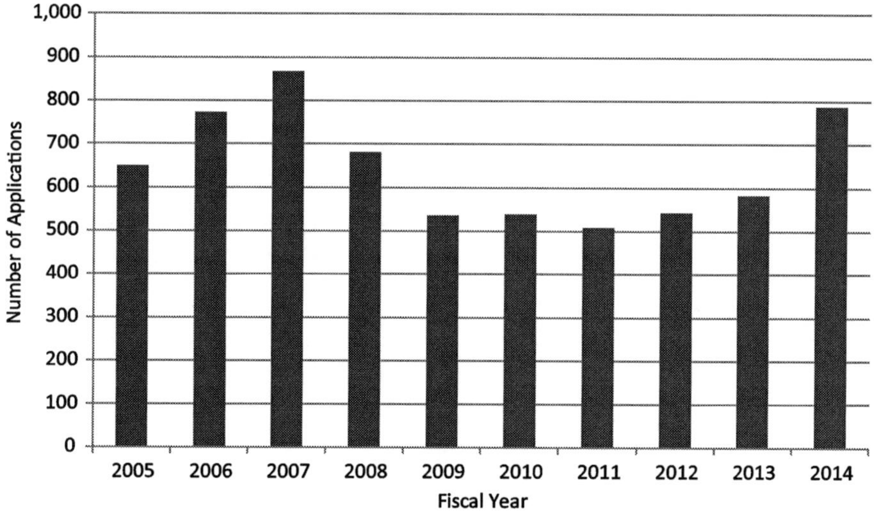

FIGURE F-7 STTR Phase I STTR applications at NIH, FY2005-2014.
SOURCE: NIH RePorter database, Table 216.

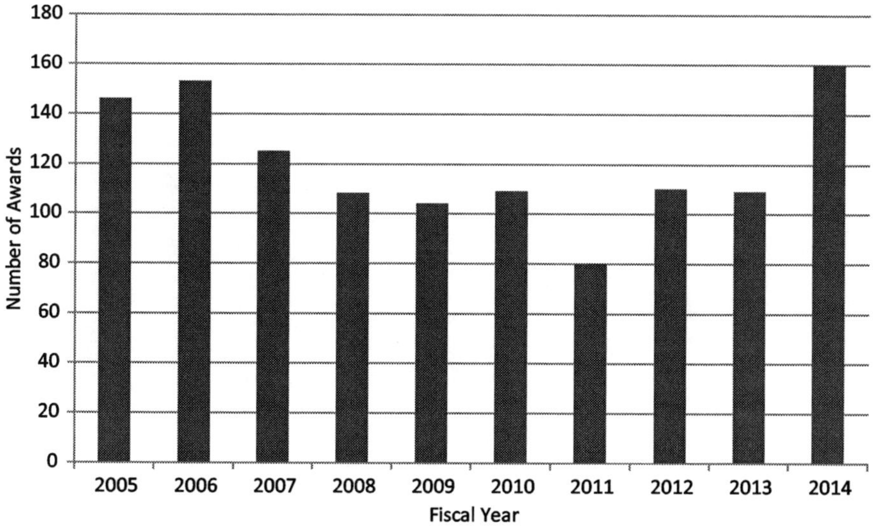

FIGURE F-8 Phase I STTR awards at NIH, FY2005-2014.
SOURCE: NIH RePorter database, Table 216.

ranging from a high of almost 50 percent in 2005 to a low of 26 percent in 2006 and 2013 (see Figure F-12). The average success rate was 35.2 percent (and that for SBIR was 38.2 percent.[5]

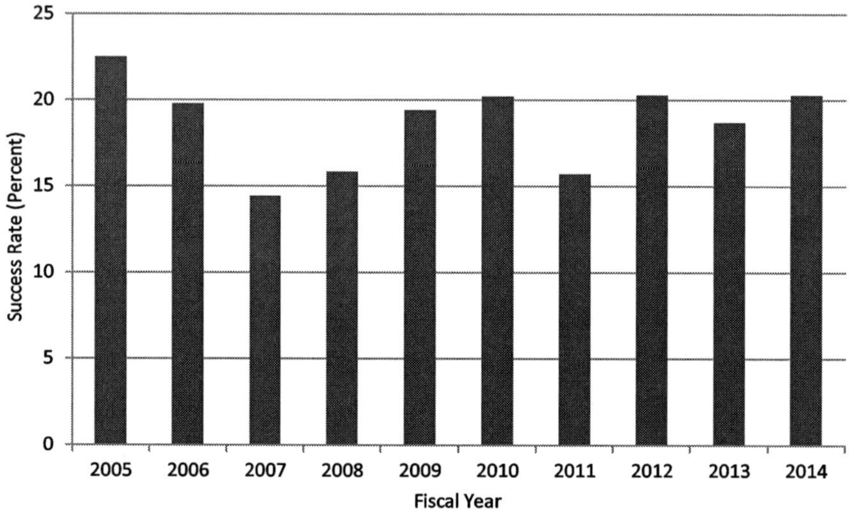

FIGURE F-9 Phase I STTR success rates at NIH, FY2005-2014.
SOURCE: NIH RePorter database, Table 216.

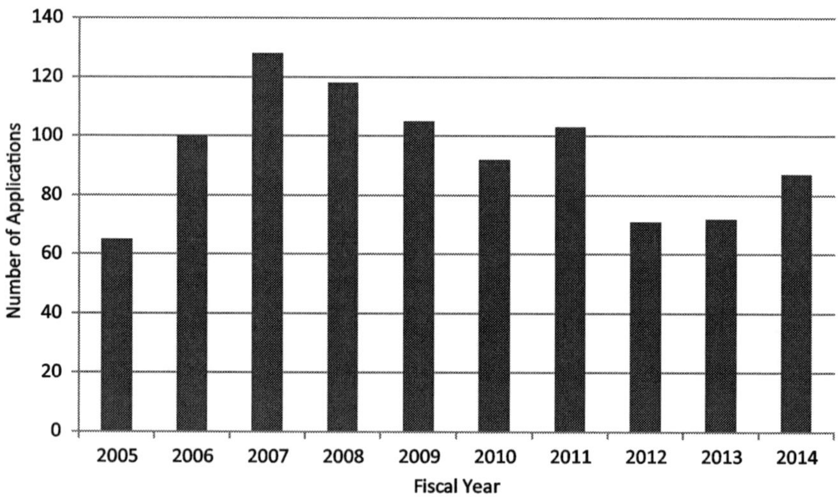

FIGURE F-10 Phase II STTR applications at NIH, FY2005-2014.
SOURCE: NIH RePorter database, Table 216.

[5]NIH RePorter database, Table 126.

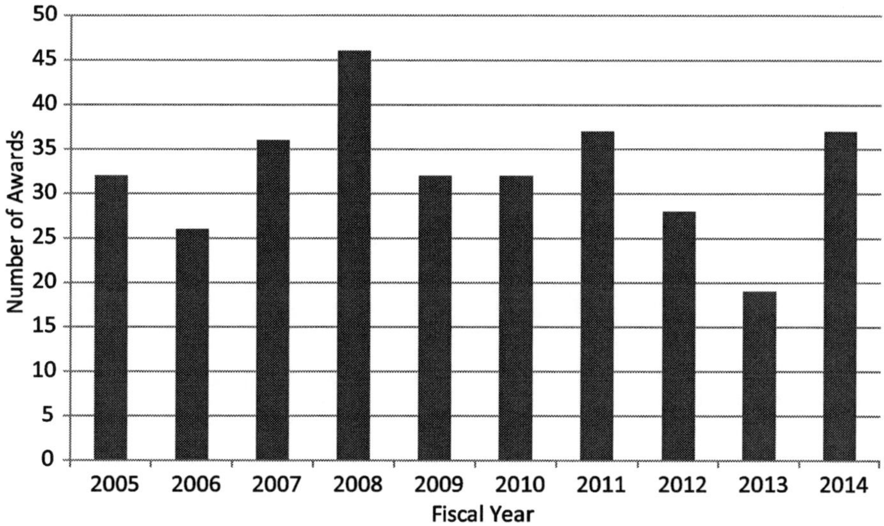

FIGURE F-11 Phase II STTR awards at NIH, FY2005-2014.
SOURCE: NIH RePorter database, Table 216.

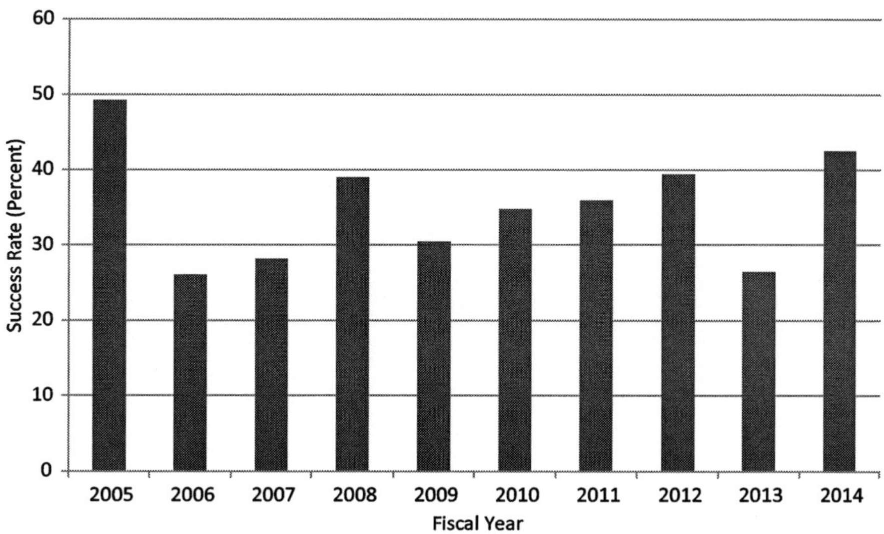

FIGURE F-12 Phase II STTR success rates at NIH, FY2005-2014.
SOURCE: NIH RePorter database, Table 216.

STTR Fast Track

NIH maintains a separate dataset covering applications and awards for Fast Track—a program through which applicants can seek funding for Phase I and Phase II through a single initial application. The number of applications was steady at just over 40 per year, with fewer in 2012 and more in 2007 and 2014 (see Figure F-13).

Since Fast Track application appears advantageous to applicants, discussions were held with NIH award recipients to assess why there are fewer applications made through Fast Track than through the traditional approach. The results suggest that many applicants believe that it is more difficult to win a Fast Track award than a Phase I award, and that NIH staff tends to discourage Fast Track applications. There is however no NIH policy to discourage Fast Track applications, according to program staff.

Although the number of Fast Track applications remained largely steady across the study period except for FY2012 and FY2014, the number of awards varied substantially in percentage terms, largely because of the small absolute numbers of awards. The difference between 4 awards (the minimum) and 12 awards (the maximum) in absolute terms is not all that large, though it represents a 300 percent increase (see Figure F-14).

High variation among the small number of Fast Track awards was also reflected in the volatility in the Fast Track success rates, which ranged from a high of 34 percent in 2012 to a low of 8 percent in 2014 (see Figure F-15).

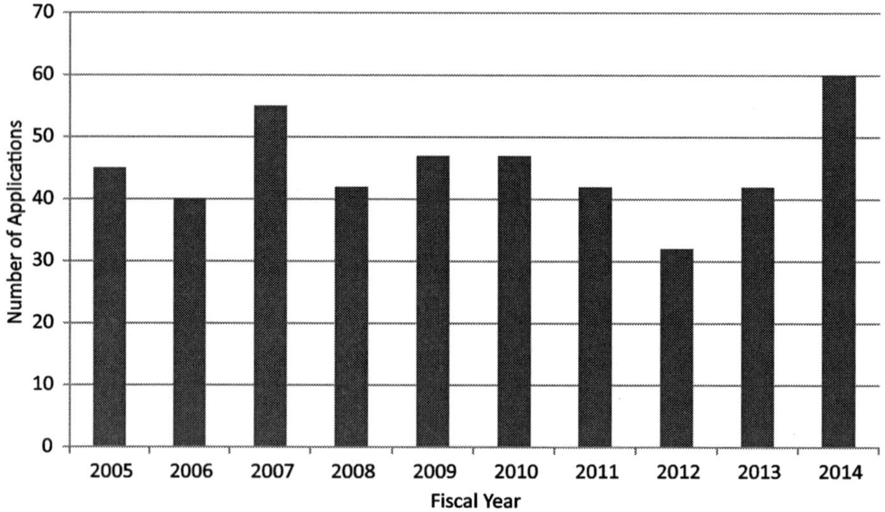

FIGURE F-13 STTR Fast Track applications at NIH, FY2005-2014.
SOURCE: NIH RePorter database, Table 216.

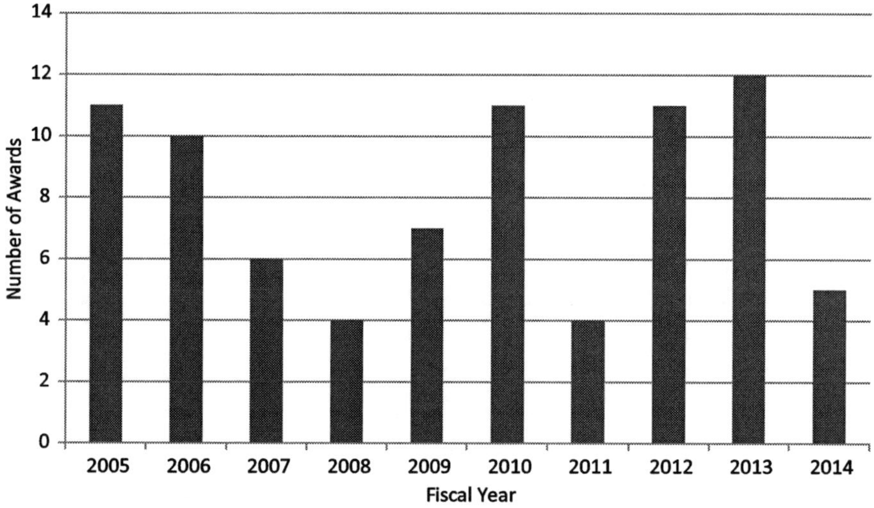

FIGURE F-14 STTR Fast Track awards at NIH, FY 2005-2014.
SOURCE: NIH RePorter database, Table 216.

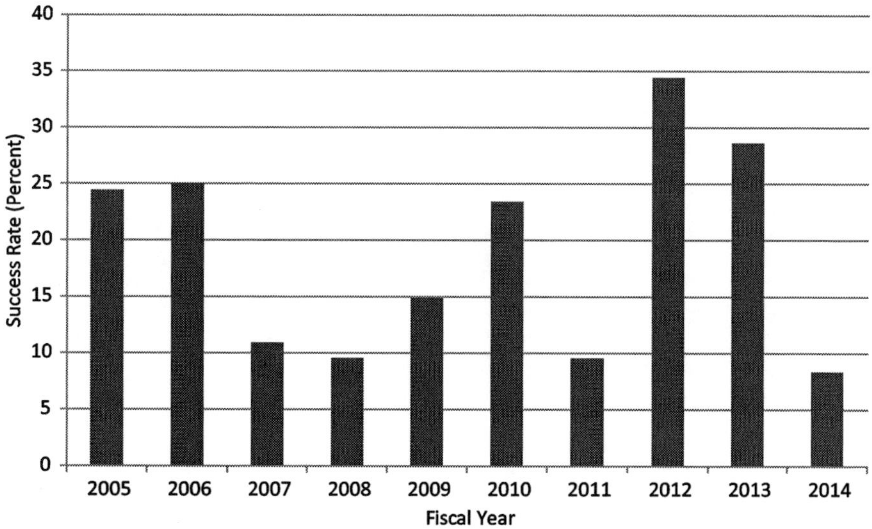

FIGURE F-15 Fast Track success rates at NIH, FY2005-2014.
SOURCE: NIH RePorter database, Table 216.

NATIONAL AERONAUTICS AND SPACE ADMINISTRATION

Phase I STTR

The number of STTR Phase I STTR awards at NASA varied sharply across the study period from a low of 101 in FY2008 to a high of almost 250 in FY2012, before declining to 101 in 2014. There were no STTR awards in 2013 due to changes in deadlines (see Figure F-16).

Over the study period, the number of Phase I STTR awards expanded from 20-25 in FY2004-2005 to 40-45 in FY2012-2014. It is unclear whether the current level will be sustained into the future, particularly in light of the recent decline in applications (see Figures F-16 and F-17).

A review of success rates reveals considerable volatility in rates, with an outlier of 39 percent in 2014 (see Figure F-18).

Phase II STTR

As with other agencies, Phase II applications at NASA are largely driven by the number of previous Phase I awards. Application levels were relatively steady during the study period, with the exception of FY2009-2011, which show the effect of ARRA funding (see Figure F-19).

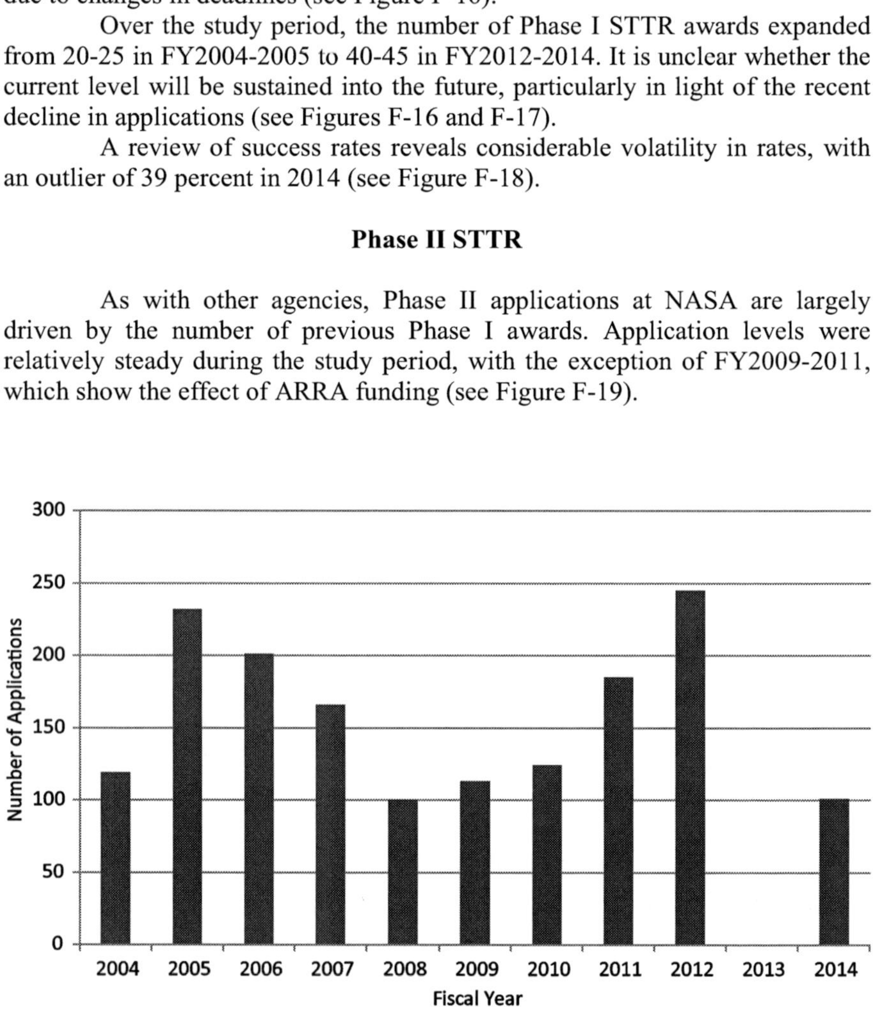

FIGURE F-16 NASA STTR Phase I STTR applications at NASA, FY2004-2014.
NOTE: Applications data for 2013 were not available.
SOURCE: Data provided by NASA.

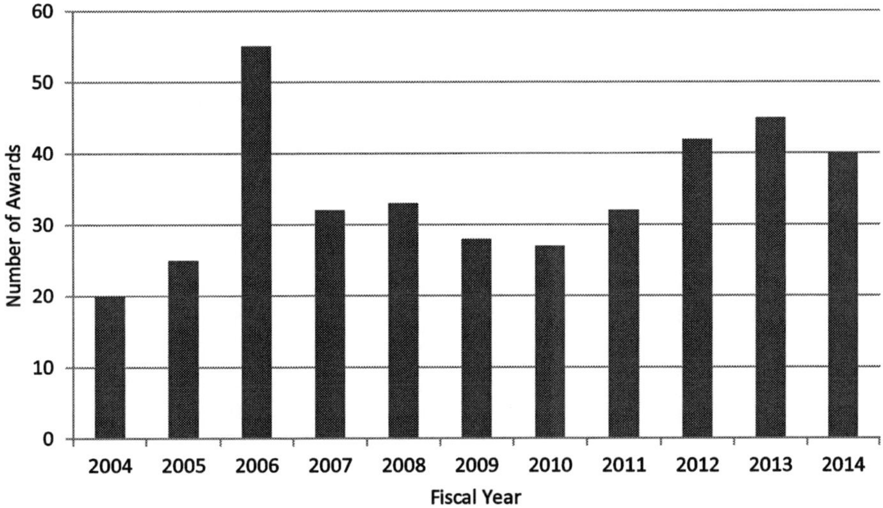

FIGURE F-17 Phase I STTR Phase Awards at NASA, FY2004-2014.
SOURCE: Data provided by NASA.

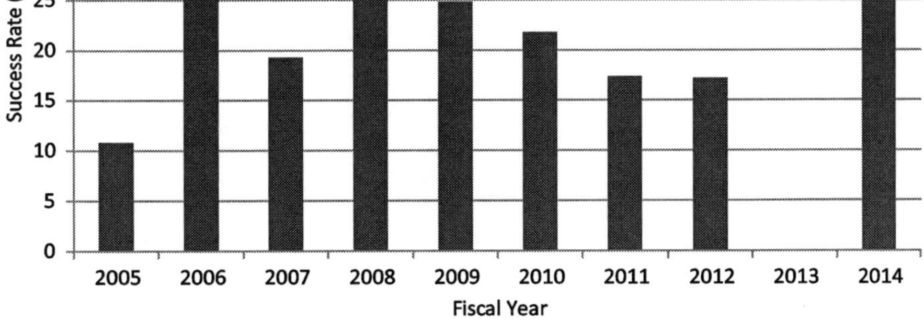

FIGURE F-18 Phase I STTR success rates at NASA, FY2004-2014.
NOTE: The absence of data for 2013 reflects the absence of applications data for that year.
SOURCE: Data provided by NASA.

Award levels doubled in 2008, from 8-9 in the preceding 3 years, and, with the exception of FY2011, remained at or above 15 awards annually through FY2013 (see Figure F-20).

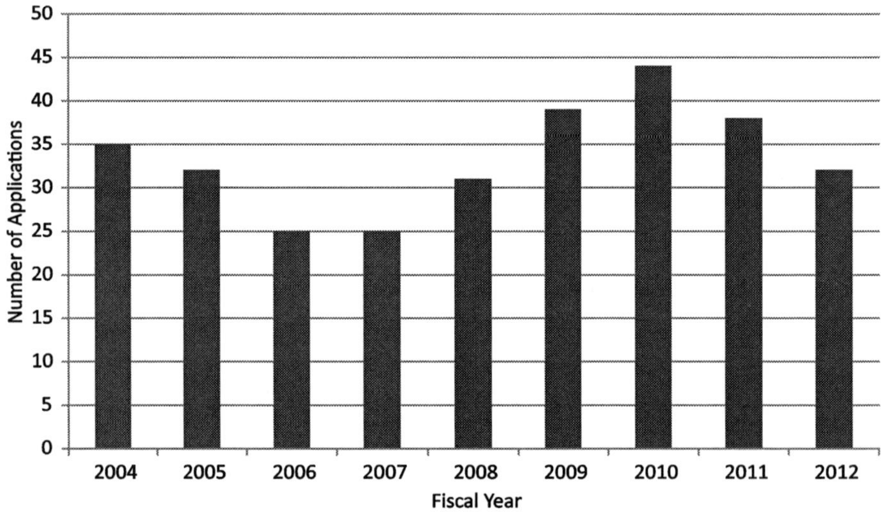

FIGURE F-19 Phase II STTR applications at NASA, FY2004-2012.
SOURCE: Data provided by NASA.

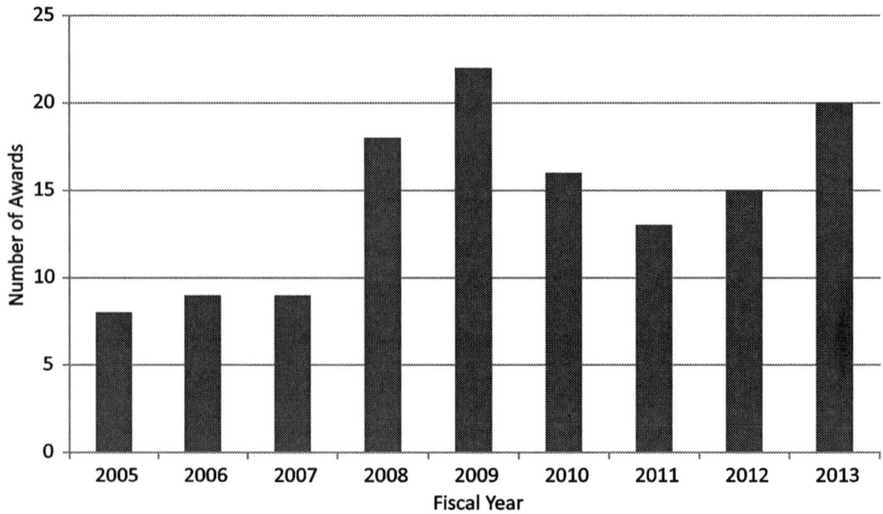

FIGURE F-20 Phase II STTR awards at NASA, FY2005-2013.
SOURCE: Data provided by NASA.

Success rates for Phase II STTR applications varied widely, from 25 percent in 2005 to greater than 50 percent in FY2008-2009, ranging between 34 and 47 percent through FY2012. The overall average from FY2005 through FY2012 was approximately 40 percent (see Figure F-21).

STTR Awards by Company

Although the 20 most prolific winners of NASA STTR of Phase I awards and of Phase II awards did not win large numbers of awards—the most prolific company won 14 Phase I awards and 7 Phase II awards during the 8-year study period—together they accounted for a substantial share of awards: 30 percent of Phase I awards and 36 percent of Phase II (see Table F-3).

TABLE F-3 STTR Awards at NASA by Company, FY2005-2012

Top 20 Phase I Company	Number of Awards	Top 20 Phase II Company	Number of Awards
CFD Research	14	CFD Research	7
Nanosonic	8	Qualtech Systems	4
Luna Innovations	7	Intelligent Fiber Optic Systems	4
Aurora Flight Sciences	6	Rolling Hills Research	4
Rolling Hills Research	6	Aurora Flight Sciences	3
Streamline Numerics	6	Applied Sensor Research & Development	3
Intelligent Fiber Optic Systems	6	Streamline Numerics	3
Applied Sensor Research & Development	5	Tetra Research	3
Tao of Systems Integration	5	TRACLabs	3
TRACLabs	5	Combustion Research and Flow Technology	3
Tetra Research	4	Luna Innovations	3
Intelligent Automation	4	Mnemonics	2
Qualtech Systems	4	Space Micro	2
Innosense	4	Sigma Space	2
American GNC	4	Balcones Technologies	2
TDA Research	4	Sustainable Innovations	2
Balcones Technologies	4	Creare	2
Plasma Processes	4	Advanced Powder Solutions	2
ZONA Technology	4	HyPerComp Engineering	2
Combustion Research and Flow Technology	3	Sikorsky Aircraft (formerly Impact Technologies)	2
Top 20 Total	107	Top 20 Total	58
Total	354	Total	160
Top 20 Percentage	30.2	Top 20 Percentage	36.3

SOURCE: Data provided by NASA.

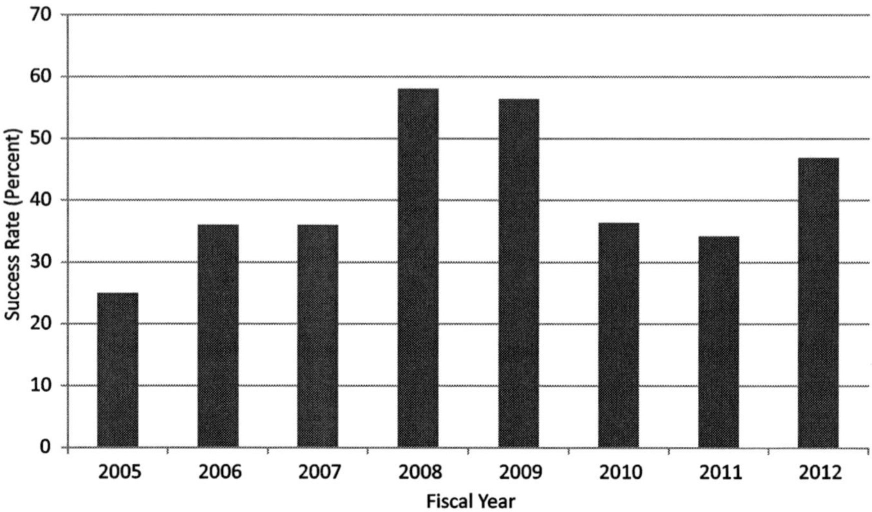

FIGURE F-21 Phase II STTR success rates at NASA, FY2005-2012.
SOURCE: Data provided by NASA.

STTR Awards by State

The distribution of awards by state may be influenced by the location of NASA research centers. States with centers may develop local innovation clusters which help SBCs to compete effectively for STTR awards. Total Phase I Awards (354) by state for the 8-year period FY2005-2012 are shown in Table F-4. Total Phase II Awards (160) by state for the 9-year period FY2004-2012 are shown in Table F-5.

NATIONAL SCIENCE FOUNDATION

Phase I STTR

The number of Phase I STTR applications at NSF varied substantially over the study period. According to NSF staff, the especially low number of applications in FY2011-2012 resulted from a pilot initiative to separate STTR and SBIR topics, creating separate topics for the former. This initiative was not popular with companies, who failed to respond, and has since been terminated. The number of applications rebounded as a result, with FY2014 seeing the highest number of applications across the study period, at almost 400 (see Figure F-22).

Phase I STTR awards at NSF tracked applications fairly well over the entire period. (See Figure F-23.)

TABLE F-4 Phase I STTR Awards at NASA by State, FY2005-2012

State	Number of Awards
AL	27
AZ	9
CA	66
CO	12
CT	10
DE	2
FL	15
GA	2
ID	1
IL	10
IN	1
KS	3
KY	2
MA	29
MD	21
MI	3
MN	3
MO	1
MS	5
MT	3
NC	1
NH	3
NJ	7
NV	2
NY	8
OH	8
OK	1
OR	1
PA	13
TN	1
TX	29
UT	4
VA	42
VT	1
WA	4
WI	4
Total	354

SOURCE: Data provided by NASA.

TABLE F-5 Phase II STTR Awards at NASA by State, FY2004-2012

State	Number of Awards
AL	11
AZ	5
CA	27
CO	4
CT	8
DE	1
FL	10
GA	1
ID	1
IL	5
KS	2
MA	10
MD	12
MI	3
MN	1
MS	2
MT	2
NH	2
NJ	3
NV	1
NY	4
OH	3
OK	1
PA	7
TX	16
UT	3
VA	12
VT	1
WA	2
Total	160

SOURCE: Data provided by NASA.

Phase I STTR success rates ranged from more than 20 percent from FY2004 to FY2009 to less than 20 percent thereafter, with the exception of FY2012, when the success rate shot up (see Figure F-24). Overall, across the

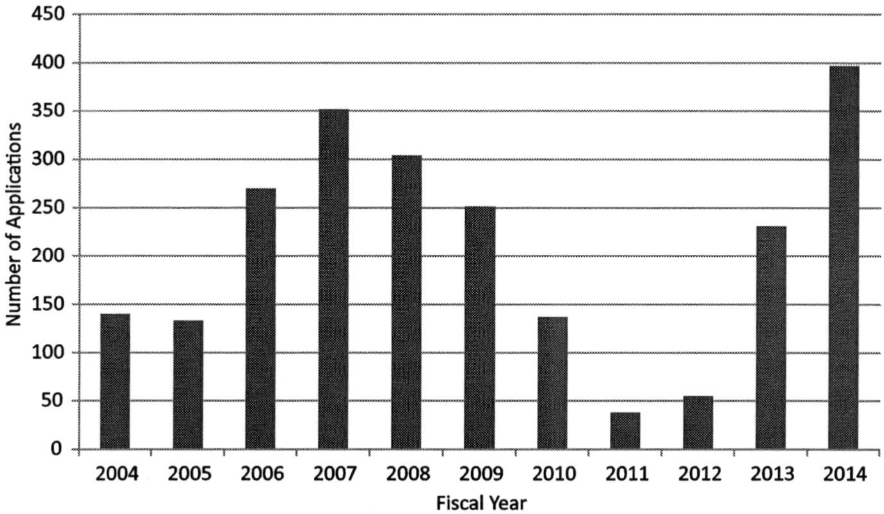

FIGURE F-22 Phase I STTR applications at NSF, FY2004-2014.
SOURCE: Data provided by NSF.

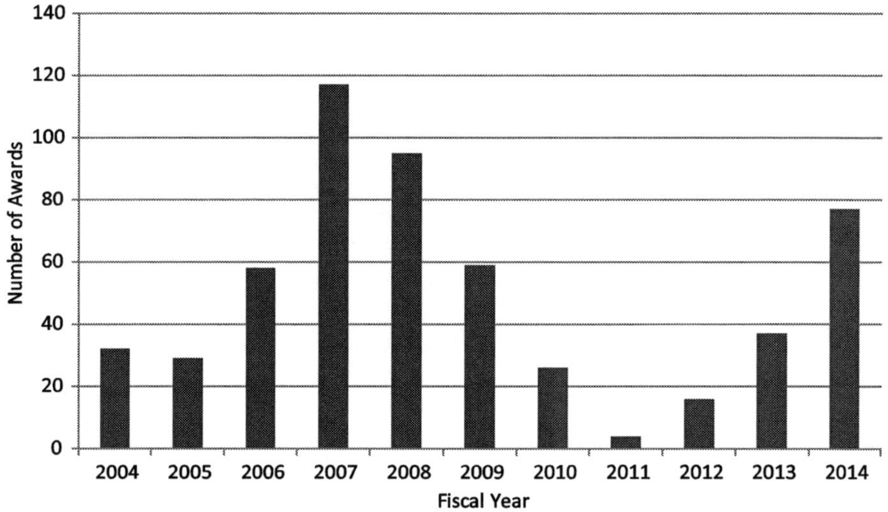

FIGURE F-23 Phase I STTR awards at NSF, FY2004-2014.
SOURCE: Data provided by NSF.

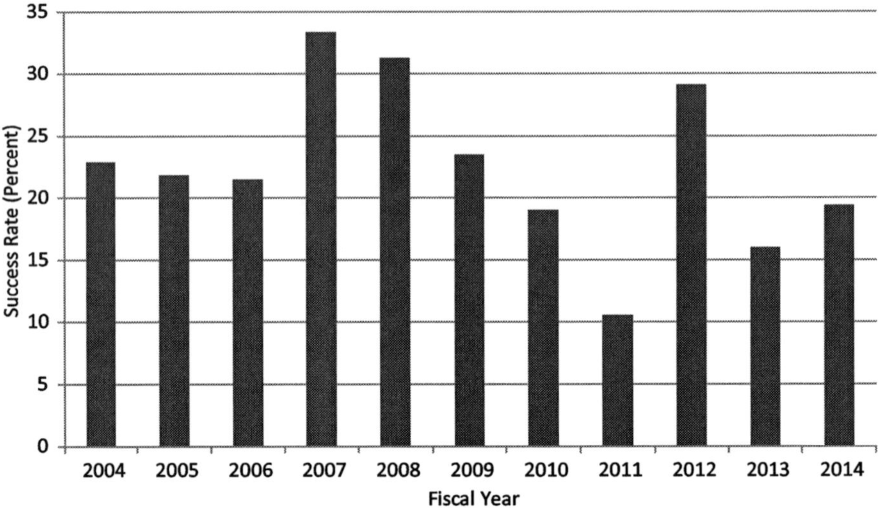

FIGURE F-24 Phase I STTR success rates at NSF, FY2004-2014.
SOURCE: Data provided by NSF.

entire study period, the success rate for Phase I STTR applications at NSF was 22.6 percent. This compares with 18.1 percent for SBIR at NSF during the same period.

Phase II STTR

As with other agencies, Phase II applications are largely driven by the number of Phase I awards made during the previous cycle. Figure F-25 shows a ramp up of Phase II STTR applications from FY2005 to FY2009, following the ramp up of Phase I awards from FY2005 to FY2007.

NSF STTR Phase II awards were particularly high in FY2008-2009, and trended down thereafter (see Figure F-26).

After the high success rates in FY2005-2006, success rates mostly varied between 30 and 40 percent (see Figure F-27). Across the 10-year study period, FY2005-2014, the average success rate for Phase II applications was 40.4 percent. The average for SBIR during the same period was 43.9 percent.

DEPARTMENT OF ENERGY

Phase I STTR

STTR applications at DoE generally remained stable during the study period. Applications averaged about 250 annually from FY2005 to FY2015 (excluding the unique ARRA solicitation in 2010) (see Figure F-28). DoE's application process is unique among the study agencies, in that applicants may apply simultaneously for STTR and SBIR funding. While this increases the number of applicants for STTR, it also provides what could be viewed as

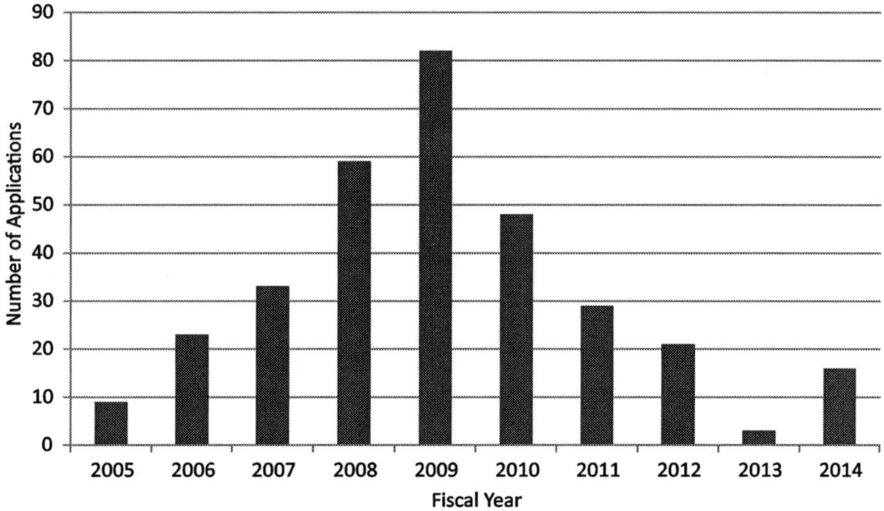

FIGURE F-25 Phase II STTR applications at NSF, FY2005-2014.
SOURCE: Data provided by NSF.

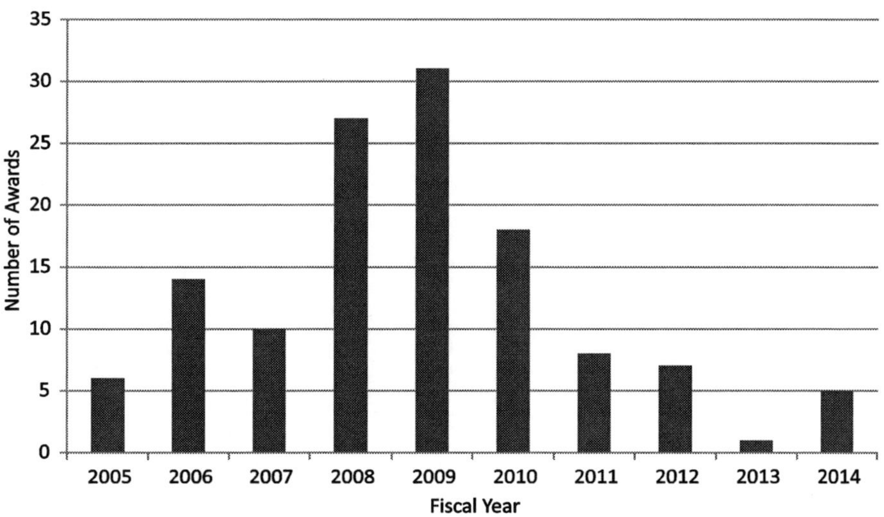

FIGURE F-26 Phase II STTR awards at NSF, FY 2005-2014.
SOURCE: Data provided by NSF.

application data that do not accurately reflect those applicants who are focused on STTR. There is no cost and no penalty to apply for both programs, as a joint application simply requires an additional checkbox entry.

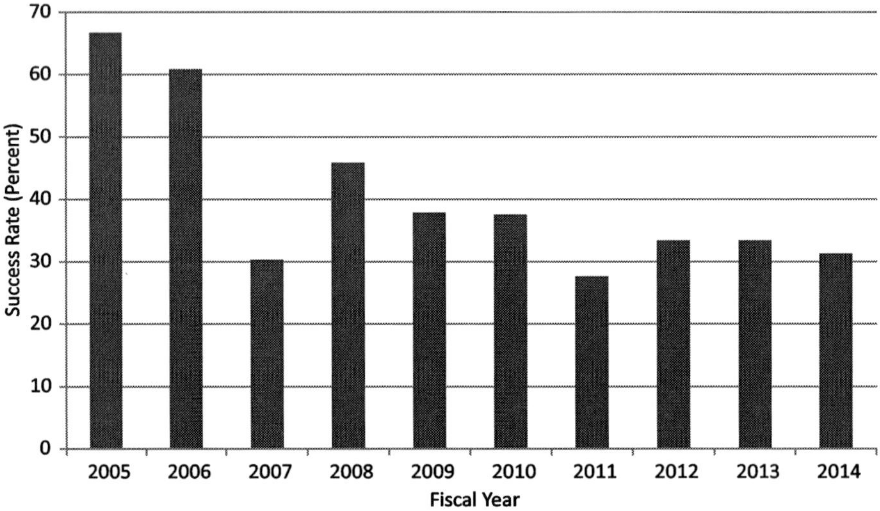

FIGURE F-27 Phase II STTR success rates at NSF, FY2005-2014.
SOURCE: Data provided by NSF.

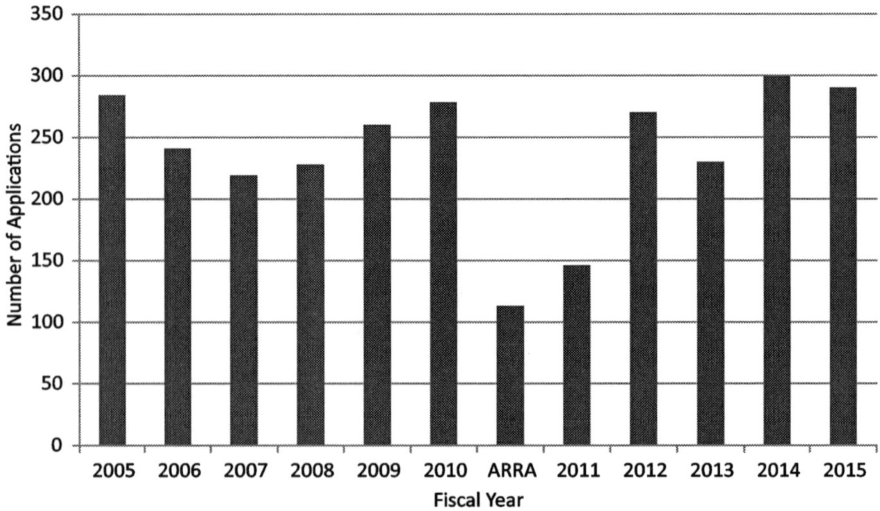

FIGURE F-28 Phase I STTR applications at DoE, FY2005-2015.
SOURCE: Data provided by DoE.

The Phase I STTR award pattern closely followed that of applications (see Figure F-29). DoE on average awarded 36 Phase I STTR awards annually (excluding the ARRA solicitation).

The close relationship between application and award patterns reflects a stable success rate (see Figure F-30). These rates are artificially low because they express Phase I STTR awards as a percentage of all applicants who checked the STTR box on the application form. Many of those were subsequently assigned to SBIR.

Phase II STTR

The steady number of Phase I STTR awards made by DoE underpins the steady flow of Phase II applications across the study period. Applications averaged 35 annually, which suggests that only 1 of the 36 Phase I projects on average per year across the study period did not apply for Phase II funding (see Figure F-31). This unusually high ratio will be explored further in an Academies report on the DoE SBIR-STTR programs, due in 2016.

Since reauthorization, Phase II STTR applications can also include Phase I SBIR winners, although DoE does not report these numbers separately. According to the DoE Program Office, it is rare for Phase I SBIR projects to apply for Phase II STTR.

During the study period, the number of Phase II awards varied from a low of 11 in the sequestration year, FY2012, to a high of 22 in FY2011. Despite

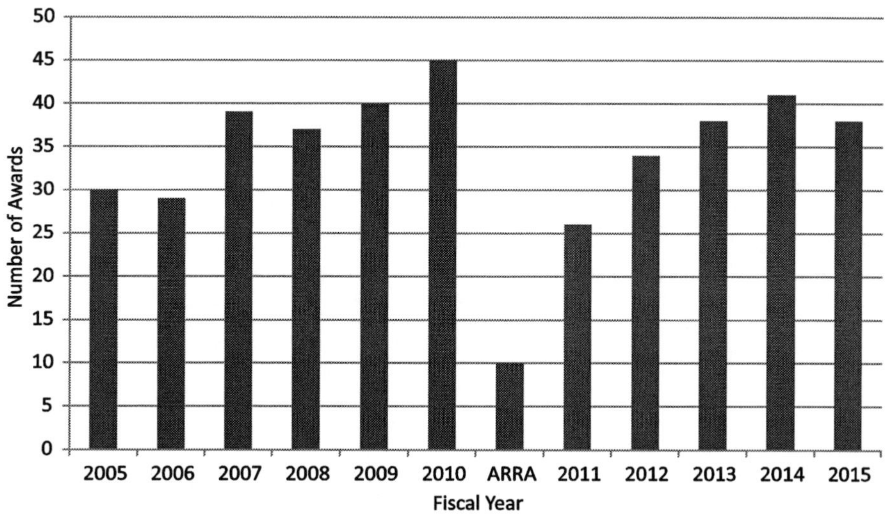

FIGURE F-29 Phase I STTR awards at DoE, FY2005-2015.
SOURCE: Data provided by DoE.

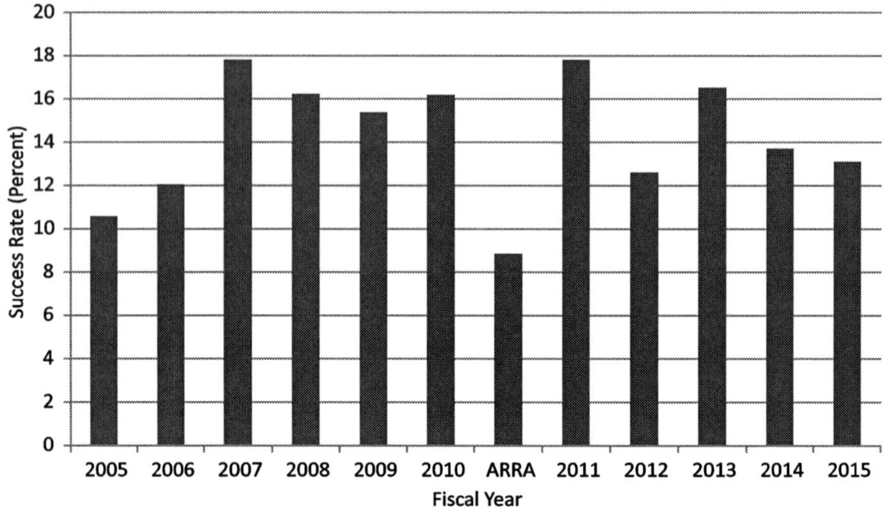

FIGURE F-30 Phase I STTR success rates at DoE, FY2005-2015.
SOURCE: Data provided by DoE.

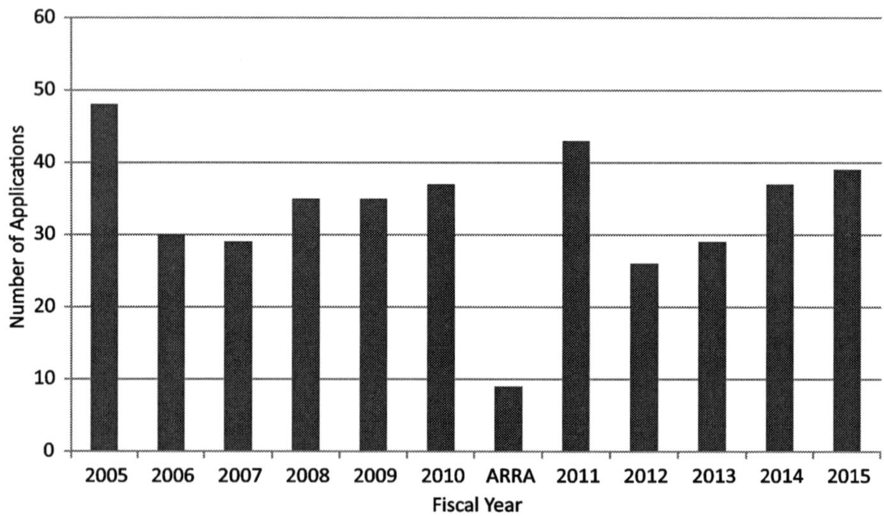

FIGURE F-31 Phase II STTR applications at DoE, FY2005-2015.
SOURCE: Data provided by DoE.

the apparent volatility, in most years DoE made between 13 and 20 awards, with an average of 16.5 (see Figure F-32, excluding the ARRA data).

The balance between awards and applications means that DoE's success rate for Phase II STTR has been very stable, with an average success rate of 47 percent (see Figure F-33).

AWARDS BY STATE

Table F-6 shows the total number of STTR Phase I awards by state for all five agencies, along with total funding and overall share of awards by state. Previous analysis by the Academies has shown that the distribution of awards is driven primarily by state population, and then by the percentage of scientists and engineers in the working population.[6] Table F-7 shows the STTR Phase II awards by state.

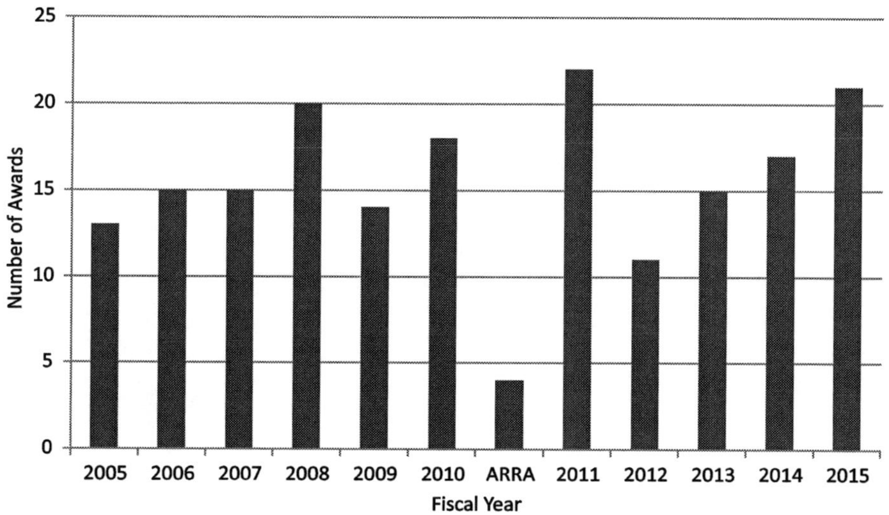

FIGURE F-32 Phase II STTR awards at DoE, FY2005-2015.
SOURCE: Data provided by DoE.

[6]National Academies, *An Assessment of the SBIR Program*, Washington, DC: The National Academies Press, 2008, Chapter 3.

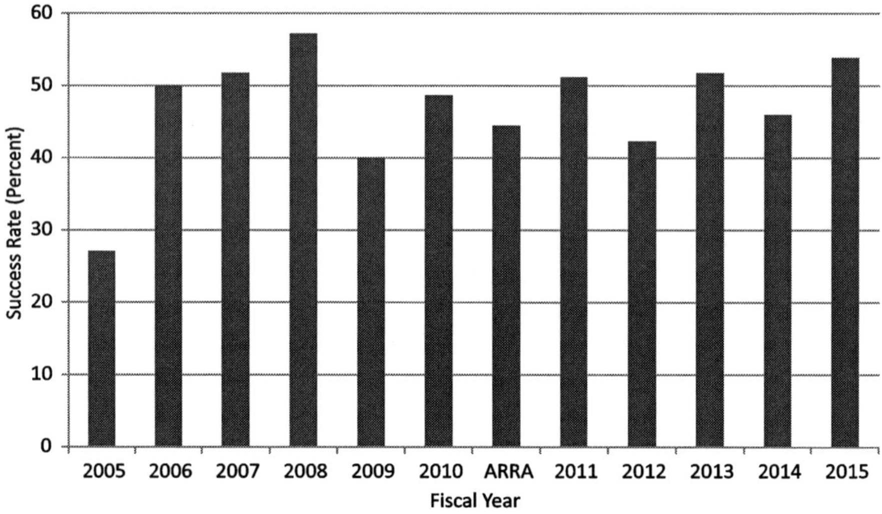

FIGURE F-33 Phase II STTR success rates at DoE, FY2005-2015.
SOURCE: Data provided by DoE.

TABLE F-6 Phase I STTR: Numbers, Amounts, and Shares of Awards by State for All Five Study Agencies, FY2005-2014

State	Number of Awards	Amount (Millions of Dollars)	Share of Awards (Percent)
AL	134	15.4	2.5
AR	34	7.0	0.6
AZ	113	14.9	2.1
CA	910	124.0	16.8
CO	220	27.3	4.1
CT	90	15.2	1.7
DC	9	1.9	0.2
DE	36	4.3	0.7
FL	154	18.7	2.8
GA	84	12.1	1.5
HI	15	1.7	0.3
IA	26	4.2	0.5
ID	8	0.8	0.1
IL	181	21.8	3.3
IN	59	8.5	1.1
KS	19	2.2	0.4
KY	57	10.9	1.1
LA	9	2.9	0.2

State	Number of Awards	Amount (Millions of Dollars)	Share of Awards (Percent)
MA	624	78.6	11.5
MD	249	35.3	4.6
ME	10	1.4	0.2
MI	130	17.9	2.4
MN	55	9.6	1.0
MO	45	6.1	0.8
MS	14	1.8	0.3
MT	29	4.8	0.5
NC	136	27.1	2.5
NE	18	3.3	0.3
NH	64	7.6	1.2
NJ	123	14.2	2.3
NM	70	7.6	1.3
NV	18	3.3	0.3
NY	241	36.5	4.4
OH	201	24.8	3.7
OK	23	4.3	0.4
OR	71	13.8	1.3
PA	184	28.4	3.4
PR	1	0.3	0.0
RI	13	2.1	0.2
SC	31	6.0	0.6
SD	6	0.8	0.1
TN	37	5.8	0.7
TX	267	34.4	4.9
UT	65	8.4	1.2
VA	355	42.2	6.5
VT	10	1.2	0.2
WA	83	12.5	1.5
WI	65	9.6	1.2
WV	12	1.2	0.2
WY	8	0.9	0.1
Total	5,416	745.6	100.0

SOURCE: SBA SBIR/STTR database.

TABLE F-7 Numbers, Amounts, and Shares of Awards by State for All Study Agencies, FY2005-2014

State	Number of Awards	Amount (Millions of Dollars	Share of Awards (Percent)
AL	37	27.6	2.7
AZ	37	26.0	2.7
CA	234	168.0	17.4
CO	66	46.1	4.9
CT	9	6.5	0.7
DC	1	0.7	0.1
DE	9	7.0	0.7
FL	36	23.4	2.7
GA	22	15.8	1.6
HI	5	3.4	0.4
IA	7	4.9	0.5
ID	3	2.0	0.2
IL	46	31.9	3.4
IN	19	14.6	1.4
KS	7	4.8	0.5
LA	1	0.6	0.1
MA	181	125.4	13.4
MD	56	37.5	4.2
ME	2	1.5	0.1
MI	24	18.0	1.8
MN	8	4.8	0.6
MO	8	5.9	0.6
MS	1	0.7	0.1
MT	4	2.7	0.3
NC	19	13.4	1.4
ND	1	0.4	0.1
NE	4	2.5	0.3
NH	15	12.8	1.1
NJ	33	22.9	2.4
NM	23	15.8	1.7
NV	2	0.8	0.1
NY	65	44.5	4.8
OH	55	39.8	4.1
OK	2	1.7	0.1
OR	4	3.9	0.3
PA	49	34.7	3.6
RI	8	6.0	0.6
SC	5	3.7	0.4
SD	1	0.7	0.1
TN	5	4.0	0.4
TX	68	45.4	5.0
UT	14	9.7	1.0
VA	103	70.3	7.6
VT	1	0.7	0.1
WA	22	14.9	1.6
WI	15	10.5	1.1
WV	6	4.2	0.4
WY	4	3.0	0.3
Total	1,347	946.1	100.0

SOURCE: SBA SBIR/STTR database.

Appendix G

Annex to Chapter 5: Quantitative Outcomes

Supplementing the analysis found in Chapter 5, this appendix reports the results of quantitative outcomes of the SBIR/STTR programs in further detail, as drawn from the 2011-2014 Academies Phase II survey. It focuses on the congressionally mandated objective for the STTR program—connecting SBCs to RIs—and also covers the other outcomes that are important to the agencies and the Congress, notably commercialization and the role of women and minorities.

CONNECTING SMALL BUSINESS CORPORATIONS (SBCs) TO RESEARCH INSTITUTIONS (RIs)

The survey asked several questions about the use of university staff and facilities on the surveyed project. The answers reveal substantial differences between SBIR and STTR respondents. Overall, 95 percent of STTR respondents reported a university connection of some kind, while only 46 percent of SBIR respondents did so.

There were also substantial differences between SBIR and STTR regarding the kind of university linkage. Thirty-two percent of STTR respondents, but only 3 percent of SBIR respondents, reported that the principal investigator (PI) was a university faculty member. STTR respondents were also more likely to report that faculty members worked on the project as a consultant (53 percent), that graduate students worked on the project (51 percent), that technology was licensed from the RIs (18 percent), and that the technology was originally developed at the RI by a project team member (29 percent) (see Table G-1). Overall, it seems clear that the university connection is much deeper and richer for STTR awards.

Respondents were also asked to identify the universities with which they worked in various capacities on this project. Although the type of help varied widely, some universities were mentioned by a number of respondents.

TABLE G-1 Links to Universities

	Percentage of Responses	
	STTR Awardees	SBIR Awardees
The PI for this project was at the time of the project an RI faculty member	32.3	2.8
The PI for this project was at the time of the project an RI adjunct faculty member	2.3	5.3
Faculty member(s) or adjunct faculty member(s) worked on this project in a role other than PI	53.0	26.0
Graduate students worked on this project	51.1	20.3
The technology for this project was licensed from an RI	18.4	6.9
The technology for this project was originally developed at an RI by one of the participants in this project	29.3	11.1
An RI was a subcontractor on this project	70.3	25.8
None of the above	4.5	53.7
BASE: TOTAL RESPONDENTS ANSWERING QUESTION	266	1,795

SOURCE: 2011-2014 Survey, Question 71.

Overall, 167 different RIs were identified from 292 projects. Those mentioned by four or more respondents are listed in Table G-2 (see Appendix D for the complete list of university mentions). Some of the names on this list are large state universities, a number of which have in recent years focused on technology transition as well as basic research. Although far from a perfect metric, we believe these data provide a preliminary indication of the connections between specific universities, university systems, and the STTR program.

STTR respondents were asked about the impact of the award on their relationship with the RI. Despite the difficulties described by some respondents and interviewees described in Chapter 4, 71 percent of respondents said that the award substantially or somewhat enhanced that relationship, while just 5 percent said that it worsened the relationship (see Table G-3). Almost three-quarters of STTR respondents indicated that they had a preexisting relationship with the RI already in place, which suggests that creating new relationships is perhaps a less important feature of the program.[1]

About eight percent of STTR respondents also received an SBIR award in which they had collaborated with an RI.[2] Among those who had received both SBIR and STTR awards, respondents were approximately evenly divided as to whether there were significant differences between SBIR and STTR.[3] This may

[1]2011-2014 Survey, Question 75.
[2]2011-2014 Survey, Question 77.
[3]2011-2014 Survey, Question 78.

TABLE G-2 University Participants Mentioned by Four or More Respondents

Research Institution	Number of Mentions
University of Florida	9
Pennsylvania State University	7
MIT	6
University of Kentucky	6
University of Massachusetts Medical School	6
Rensselaer Polytechnic Institute	5
University of Colorado-Boulder	5
Vanderbilt University	5
Boston University	4
Georgia Tech	4
Indiana University	4
Johns Hopkins University	4
Lawrence Berkeley National Laboratory	4
National High Magnetic Field Laboratory (NHMFL)	4
Texas A&M University	4
University of California Berkeley	4
University of Illinois-Chicago	4
University of Minnesota	4

SOURCE: 2011-2014 Survey, Question 72.

simply reflect the differing practices of the awarding agency—as described in Chapter 2, for NIH, DoE, and NSF there are effectively no programmatic differences.

Almost one-half also indicated that STTR was more difficult to manage (see Figure G-1). Only 3 percent thought STTR was easier to manage. This corresponds with the views expressed in Chapter 4.

Survey respondents generally agreed that the share of funding going to the RI should not be increased—only about 18 percent agreed with this concept (see Table G-4).

It is also worth observing that company founders are closely connected to the universities. For almost 80 percent of STTR companies in the sample, at least one founder had an academic background (see Table G-5).

BOX G-1
Workshop on Improving University-SBIR/STTR Linkages

A workshop convened on February 5, 2014, at the committee considered a range of issues concerning universities and the SBIR/STTR programs.[a] Participants at this workshop discussed a range of topics including

- Improving linkages between the SBIR/STTR programs at agencies and the universities,
- Aligning with university accelerator initiatives,
- Supporting improved links between state and local innovation and entrepreneurship programs and the universities, and
- Supporting shifts in culture at universities to incentivize faculty to pursue SBIR/STTR funding.[b]

[a]See http://sites.nationalacademies.org/PGA/step/sbir/PGA_086819.htm.
[b]These issues and others related to the SBIR/STTR programs and universities will be addressed in detail in a forthcoming Academy report on the NASA SBIR program.

TABLE G-3 STTR Impacts on SBC-RI Relationships

	Percentage of Responses
Substantially enhanced it	37.8
Somewhat enhanced it	33.2
Made no real difference	23.7
Made it somewhat worse	4.6
Made it substantially worse	0.8
BASE: STTR AWARD RECIPIENTS	262

SOURCE: 2011-2014 Survey, Question 74.

While 56 percent of company founders were previously employed at other private companies, 59 percent of respondents reported at least one founder previously employed at an RI or a National Lab (see Table G-6).

THE FOCUS ON COMMERCIALIZATION OUTCOMES

Although the statutory goal for the STTR program is to enhance linkages between RIs and SBCs, as a practical matter, STTR programs are still largely judged by their success in commercializing technologies.[4] Moreover, given that commercialization is among the more measurable outcomes of the

[4]SBA, Section 1.(c), STTR Policy Directive, October 18, 2012, p. 3.

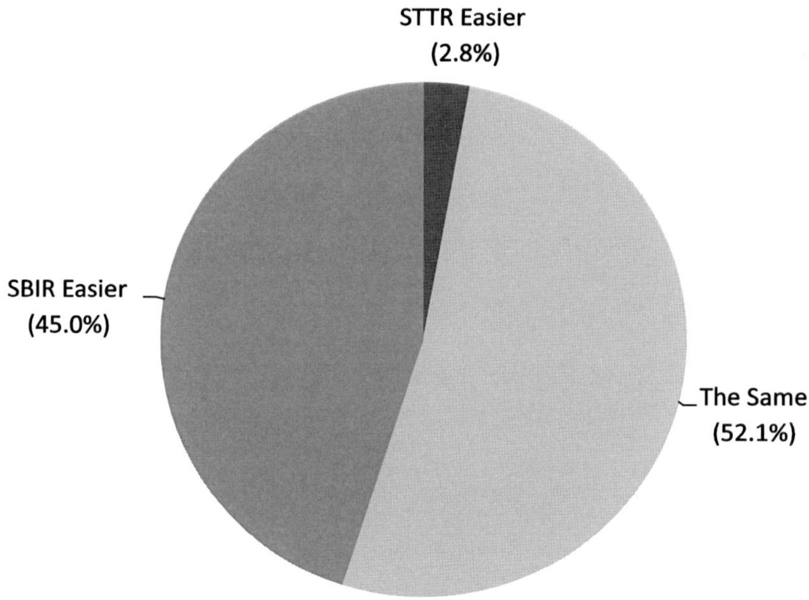

FIGURE G-1 Comparative ease of managing SBIR and STTR.
SOURCE: 2011-2014 Survey, Question 80. N=211.

TABLE G-4 Increased Share of Funding for Research Institutions

	Percentage of Responses
Strongly agree	8.6
Somewhat agree	9.0
Neither agree nor disagree	31.6
Somewhat disagree	23.7
Strongly disagree	27.1
BASE: STTR AWARD RECIPIENTS	266

SOURCE: 2011-2014 Survey.

SBIR/STTR programs, it has become a primary benchmark for program performance. The focus on commercialization, however, should not be allowed to obscure the requirement that the program meet all congressionally mandated objectives. This appendix provides additional details of the commercial outcomes of the SBIR/STTR programs, as well as quantitative outcome measures related to expanding the U.S. science and engineering base.

TABLE G-5 Number of Academic Founders

	Percentage of Company Responses	
	STTR Awardees	SBIR Awardees
0	21.2	27.7
1	36.0	42.2
2	25.3	18.1
3	11.8	6.7
4	3.0	3.6
5 or more	2.6	1.8
Mean	1.50	1.24
Median	1.00	1.00
BASE: TOTAL RESPONDENTS ANSWERING QUESTION	168	1,017

NOTE: Because multiple responses were received from some companies, responses here are weighted to provide the average response per company.
SOURCE: 2011-2014 Survey, Question 5.4.

TABLE G-6 Prior Employment of Founders

	Percentage of Company Responses	
	STTR Awardees	SBIR Awardees
Other private company	55.9	64.8
Research institution	57.3	40.1
Government	7.2	6.5
FFRDCs or National Labs	1.4	0.9
Other	3.9	8.3
BASE: TOTAL COMPANIES ANSWERING QUESTION	173	1,039

NOTE: Because multiple responses were received from some companies, responses here are weighted to provide the average response per company.
SOURCE: 2011-2014 Survey, Question 6.

SOURCES OF DATA

All the major SBIR agencies either have in place data collection systems related to outcomes from SBIR/STTR awards or are putting such systems in place.

- DoD continues to maintain the Company Commercialization Record (CCR), which requires all companies applying for DoD SBIR/STTR funding to update outcomes for all prior awards.

- NSF utilizes a consultant to undertake phone interviews with recipients at set times certain years after the end of the award.
- DoE maintains internal tracking of award outcomes using its own metrics and methodologies.
- NASA developed a tracking module as part of its Electronic Hand Book, which has in recent years been used to begin collecting outcomes data.
- NIH is seeking to work closely with SBA as the latter develops a tracking system, and it is also working to improve its own data collection and analytics system beyond the current limited scale and scope.

However, these approaches do not generate comparable data. The agencies use different mechanisms for collecting data, ask companies different questions, and provide different levels of data access to the Academies (e.g., DoD provided complete access to the CCR data set, while NSF declined to provide its outcomes data on privacy grounds). Because of the lack of comparable data across all agencies, the quantitative data presented in this chapter are derived from the Academies 2011-2014 Phase II survey of award recipients. However, these data are descriptive only and should be regarded as providing insights into outcomes rather than definitive conclusions.[5]

The 2011-2014 Survey fielded by the Academies was based primarily on the previous 2005 Academies survey, with some additions and modifications. The survey was deployed twice: to DoD, NSF, and NASA award recipients in 2011 (covering awards made in FY1998-2007 inclusive) and to NIH and DoE award recipients in 2014 (covering awards made in FY2001-2010 inclusive). These award time frames are referred to as the "study period" throughout this report.

The 2011-2014 Survey was sent to all PIs who received an SBIR/STTR or STTR Phase II award from one of the five study agencies during the study period. In an effort to improve response rates for NIH and DoE, when a PI could not be reached, the survey was sent to alternate company contacts at the targeted companies (which generated approximately 100 additional responses).

Appendix A provides a detailed discussion of the survey methodology, including response rates and potential survey bias. The text below provides a full series of tables summarizing Phase II responses for STTR and SBIR recipients. The 2011-2014 Survey is reproduced in Appendix C.

---

[5]The committee previously sought to develop statistical comparisons with similar companies in similar sectors at similar stages of development, but these efforts were eventually abandoned as unworkable. See Appendix A for a discussion of this effort. A full description of the methodology employed for this survey and the resulting analysis is also provided in Appendix A.

COMMERCIALIZATION

The agencies vary considerably in their use of STTR, which is reflected in their differing views of commercialization, described in more detail in the series of Academies reports on the individual agencies. Nonetheless, there are data and related metrics that help to provide useful insights into commercialization across all the agencies.

Defining "Commercialization"

Several important conceptual challenges emerge when seeking to define "commercialization" for the purposes of the SBIR and STTR programs. Like many apparently simple concepts, commercialization becomes progressively more difficult and complex as it is subjected to further scrutiny. For example:

- Should commercialization include just sales or other kinds of revenue as well, such as licensing fees and funding for further development?
- Should commercialization include only certain kinds of sales—excluding, for example, sales to government agencies?
- What is the appropriate benchmark for sales? Is it any sales whatsoever, sufficient sales to cover the costs of awards, sales that lead to breaking even on a project, or sales that reflect a commercial level of success and viability? The latter at least would likely be different for each project in each company.
- Should commercialization include sales by licensees, which may be many multiples of royalty revenues provided to STTR recipients, but are more difficult to track and to assign causality to specific STTR awards?

For the purposes of this study, we deployed a broad net to capture a range of potentially useful data. We include all revenues related to the funded project. We count all sales to any customer as a sale. We exclude all revenues from licensees because we do not believe the data reported by the original technology developer are likely to be sufficiently accurate, even though in some important cases licensees do use SBIR/STTR-developed technologies to build large-scale businesses. This approach is identical to that used in all previous Academies studies of the SBIR program. Once acquired, these data can be analyzed in a variety of ways to provide multiple insights into this complex topic.[6]

[6]For an overview of the commercialization metrics and survey used in this study, see Appendix A.

Project-related Revenues

Perhaps the single most used metric for assessing SBIR-type programs is project-related revenues. Although previous Academy reports have cautioned against overuse of this metric—warnings that are reflected in the wide range of metrics adopted for use in the current assessment—project-related revenues are still an important metric.[7]

Reaching the market. The first survey question in this area concerns reaching the market: Did the project generate any sales, and if not, are sales expected (a necessary question given the long cycle time of some projects)? Responses are summarized in Table G-7. Slightly less than 40 percent of STTR projects reported some sales or licensing revenues, compared with 49 percent of SBIR projects. A further 28 percent of STTR respondents expected sales in the future, compared with 24 percent of SBIR respondents. This suggests that STTR projects may on average have a longer cycle.

Amount of Sales and Licensing Revenues. The percentage of projects reaching the market is an important metric, but it is not sufficient. It is also important to understand the distribution of sales. The survey asked those respondents who reported some sales of the technology developed for the project to report the amount of sales, divided into dollar ranges (see Table G-8).

TABLE G-7 SBIR/STTR Sales

	Percentage of Responses	
	STTR Awardees	SBIR Awardees
No sales to date	60.1	50.7
No sales to date nor are sales expected	32.1	26.2
No sales to date, but sales are expected	28.0	24.5
Any sales to date	39.9	49.3
Sales of product(s)	30.6	36.1
Sales of process(es)	1.8	4.6
Sales of services(s)	14.4	22.0
Other sales (e.g., rights to technology, licensing, etc.)	6.6	7.1
BASE: TOTAL RESPONDENTS ANSWERING QUESTION	271	1,856

NOTE: Respondents could select multiple types of sales for a single project, so percentages for types of sales do not sum to "Any sales to date."
SOURCE: 2011-2014 Survey, Question 32.

[7]Similar warnings can be found in the 2009 report on the NIH SBIR program by The Academies—National Research Council, An Assessment of the SBIR Program at the National Institutes of Health, Washington, DC: The National Academies Press, p. 81.

TABLE G-8 Distribution of Total Sales, by Range and Phase

	Percentage of Responses	
	STTR Awardees	SBIR Awardees
None ($0)	3.1	3.0
Under $100,000	25.5	27.0
$100,000-$499,999	33.7	27.0
$500,000-$999,999	17.3	13.1
$1,000,000-$4,999,999	14.3	21.1
$5,000,000-$9,999,999	4.1	4.6
$10,000,000-$19,999,999	1.0	2.0
$20,000,000-$49,999,999	1.0	1.4
$50,000,000 or more		0.7
Mean (Thousands of Dollars)	1,488.8	2,479.5
Median (Thousands of Dollars)	300	300
BASE: ANY SALES RESULTING FROM THE PROJECT	98	862

SOURCE: 2011-2014 Survey, Question 34.

Most respondents reported sales at the lower end of the ranges: 62 percent of STTR respondents reported revenues of less than $500,000, compared with 57 percent of SBIR respondents. One percent reported revenues of at least $20 million. The substantially different means indicate important positive outliers in the SBIR group but not in the STTR group. There was no difference between the groups in the medians.

Markets by Sector

The survey asked respondents about the market sectors in which sales were made. Overall, 40 percent of respondents identified the domestic private sector, followed by DoD and DoD contractors combined (25 percent), and export markets (17 percent) (see Table G-9). No other single sector reached 5 percent. The market by sector varies substantially by agency—for example, DoD projects largely focus on DoD markets, while NIH and NSF projects largely focus on the domestic private sector.

EMPLOYMENT

As with prior surveys, respondents were asked both about the size of the company at the time of the award and the current size, in terms of number of employees. Table G-10 shows that the median and mean of employees for STTR

TABLE G-9 Markets for STTR and SBIR Products and Services

	Percentage of Total Sales	
	STTR Awardees	SBIR Awardees
Domestic private sector	39.8	39.0
Export markets	16.6	11.5
Department of Defense (DoD)	14.4	17.4
NASA	2.0	2.7
Prime contractors for DoD	10.4	10.5
Prime contractors for NASA	0.5	0.9
Agency that awarded the Phase II (if not NASA or DoD)	1.8	1.2
Other federal agencies	3.5	4.2
State or local governments	0.4	2.5
Other	10.6	10.1
BASE: ANY SALES RESULTING FROM THE PROJECT	99	889

NOTE: Respondents were asked to provide a percentage breakdown by market. The table shows the mean of responses for each category.
SOURCE: 2011-2014 Survey, Question 35.

TABLE G-10 Number of Employees at Time of Award

	STTR (Percentage of Companies Responding)
0	1.2
1	6.3
2	10.1
3 or 4	19.6
5 to 9	25.6
10 to 19	15.2
20 to 49	13.0
50 to 99	4.9
100 or more	4.0
Mean	19
Median	6
BASE: TOTAL RESPONDENTS ANSWERING QUESTION	163

NOTE: For questions where company (rather than project) responses are reported, SBIR responses are not included because in many cases STTR winners also received SBIR awards and vice versa.
SOURCE: 2011-2014 Survey, Question 14.1.

companies were 6 and 19, respectively, at the time of the award. Both numbers are smaller than for SBIR companies, of which some had 100 or more employees and fewer had 0-2 employees.

For current employees, STTR respondents reported that the median number grew from 6 to 7, The mean size expanded for from 19 to 50. Seven percent of STTR companies reported at least 100 employees (see Table G-11). For STTR, current employment is still concentrated in the smallest firms, with 57 percent of respondents indicating current company size as 9 or less.

ADDITIONAL INVESTMENT

The ability of projects and companies to attract additional investment has traditionally been a defining metric for SBIR/STTR commercialization outcomes.[8] There has also been interest in the sources of additional funding for high-tech innovation because the United States has historically been at the forefront of venture capital and angel investment.

Seventy-one percent of STTR respondents indicated that their company had received additional investment in the technology related to the surveyed project. As with prior surveys, there is substantial skew with regard to the

TABLE G-11 Number of Employees at Time of Survey

	STTR (Percentage of Companies Responding)
0	10.8
1	5.0
2	8.0
3 or 4	11.9
5 to 9	21.1
10 to 19	18.0
20 to 49	11.5
50 to 99	6.6
100 or more	7.3
Mean	50
Median	7
BASE: TOTAL COMPANIES ANSWERING QUESTION	161

SOURCE: 2011-2014 Survey, Question 14.2.

[8]See National Research Council, *An Assessment of the SBIR Program*, Washington, DC: The National Academies Press, 2008. Effective July 1, 2015, the institution is called the National Academies of Sciences, Engineering, and Medicine. References in this report to the National Research Council or NRC are used in an historic context identifying programs prior to July 1.

amount of additional funding received: 29 percent did not raise any funding, and 46 percent of STTR respondents reported receiving less than $100,000, while slightly less than 1 percent reported more than $20 million (see Table G-12).

Of those projects that received additional funding, 54 percent reported funding from internal sources, and 46 percent from non-SBIR/STTR federal sources. Six percent of STTR respondents indicated that they had received venture capital funding, and 3 percent received funding from angel and other private equity investors. Fifteen percent reported strategic investments from partners (see Table G-13).

COMPANY-LEVEL COMMERCIALIZATION THROUGH MERGERS AND ACQUISITIONS

SBIR/STTR firms often commercialize their technology through mergers or other company-level activities. Sixty-seven percent of STTR respondents indicated that their companies had not been acquired, had not implemented or planned an initial public offering (IPO), and had not established a spin-off. Twelve percent reported that their company had spun off one or more new companies, 16 percent that they had entered into a strategic partnership, and

TABLE G-12 Additional Funding by Phase and Amount

	Percentage of Responses	
	STTR Awardees	SBIR Awardees
None ($0)	29.2	31.3
Under $100,000	46.2	47.1
$100,000-$499,999	9.5	8.5
$500,000-$999,999	6.1	4.3
$1,000,000-$4,999,999	7.2	6.2
$5,000,000-$9,999,999	0.8	1.1
$10,000,000-$19,999,999	0.4	0.7
$20,000,000-$49,999,999	0.8	0.6
$50,000,000 or more		0.2
Mean (Thousands of Dollars)	691.7	832.1
Median (Thousands of Dollars)	50	50
BASE: TOTAL RESPONDENTS ANSWERING QUESTION	264	1,827

SOURCE: 2011-2014 Survey, Question 30.

TABLE G-13 Distribution of Additional Investment Funding by Source of Funds

	Percentage of Responses	
	STTR Awardees	SBIR Awardees
Non-SBIR/STTR federal funds	45.5	39.5
Private Investment: U.S. Sources	30.2	35.6
Venture capital (VC)	6.3	5.4
U.S. angel funding or other private equity investment (not VC)	3.3	8.4
Friends and family	5.3	4.3
Strategic investors/partners	15.3	13.1
Other sources	11.1	15.3
Foreign Investment	2.6	4.0
Other External Sources	*32.3*	*14.2*
State or local governments	16.9	10.7
Research institutions (such as colleges, universities or medical centers)	17.5	4.4
Foundations	3.2	1.0
Internal Sources	*54.0*	*69.9*
Your own company (Including money you have borrowed)	45.5	62.1
Personal funds	14.3	17.3
BASE: TOTAL RESPONDENTS ANSWERING QUESTION	189	1,239

NOTE: Responses for subcategories do not total to categories because more than one response was permitted.
SOURCE: 2011-2014 Survey, Question 31.

12 percent that the awardee company had been acquired or merged with another company. SBIR firms were more likely to establish a spin-off company, but otherwise responses for the two groups were similar (see Table G-14).

COMMERCIALIZATION TRAINING AND MARKETING

Federal agencies have in recent years provided more commercialization training for SBIR awardees. In some cases this training has been mandatory. Table G-15 indicates that 30 percent of STTR awardees received agency-sponsored training.

Of those who participated, 29 percent of STTR respondents thought that the training was valuable or extremely valuable. Conversely, about one-quarter of participants considered it not very valuable or not at all valuable.

TABLE G-14 Company-Level Changes

	Percentage of Responses	
	STTR Awardees	SBIR Awardees
Established one or more spin-off companies	11.5	18.7
Entered into strategic partnership with major industry player	16.4	16.1
Been acquired by/merged with another firm	11.6	8.6
Made an initial public offering	1.9	1.9
Planning to make an initial public offering in the next two years	1.2	1.9
None of the above	67.2	63.0
BASE: TOTAL RESPONDENTS ANSWERING QUESTION	167	1,019

NOTE: Because multiple responses were received from some companies, responses here are weighted to provide the average response per company.
SOURCE: 2011-2014 Survey, Question 11.

TABLE G-15 Participation in Commercialization Training (percentage of responses)

	Percentage of Responses	
	STTR Awardees	SBIR Awardees
Yes	30.0	36.1
No	70.0	63.9
BASE: TOTAL RESPONDENTS ANSWERING QUESTION	267	1,809

SOURCE: 2011-2014 Survey, Question 49.

SBIR recipients were more enthusiastic about the training (see Table G-16), which suggests that the training may not have been geared effectively to the more academic participants who entered the program through STTR.

The survey also asked respondents about new support opportunities generated through the reauthorization legislation, which permits Phase II companies to spend up to $10,000 each on marketing support, as an alternative to the commercialization training offered through agency-sponsored providers. Table G-17 shows that about one-half of STTR respondents would prefer to use the funding on their own, while 37 percent would prefer to continue using agency providers.

TABLE G-16 Effectiveness of Commercialization Training

	Percentage of Responses	
	STTR Awardees	SBIR Awardees
Extremely valuable	11.4	10.6
Very valuable	17.1	28.2
Somewhat valuable	45.7	36.3
Not very valuable	20.0	19.8
Not at all valuable	5.7	5.1
BASE: ACCEPTED COMMERCIALIZATION ASSISTANCE IN CONNECTION WITH AWARD	35	273

SOURCE: 2011-2014 Survey, Question 51.

KNOWLEDGE EFFECTS

One of the congressionally mandated objectives for the SBIR/STTR programs is to "stimulate technological innovation," which is often equated with patenting activity. However, in the context of small business, this standard metric of innovation does not capture the entire story: patenting is important, but it is also expensive, and SBIR/STTR funds cannot legally be used for this purpose. Many companies interviewed for this (and for previous reports by the Academies on SBIR) indicated that they preferred to keep their technology secret or to rely on first-mover advantages and other market-based leverage to defend their technologies.

However, standard metrics provide at least a starting point for quantitative analysis. Consequently, the survey addressed several intellectual property (IP)-related metrics: patents, trademarks, copyrights, and peer-reviewed papers.[9]

Patents

Patents are to some degree the life blood of high-tech firms. Overall, about one-half of STTR respondents (and two-thirds of SBIR respondents) claimed to have been awarded at least one patent related to any SBIR-funded technology; 5.5 percent of STTR respondents reported at least 10 related patents (see Table G-18). SBIR companies on average (mean) reported more than twice as many patents as did STTR companies.

[9]The values of these knowledge repositories vary. Any unique item, painting, photo, or music score can be copyrighted for a modest fee. Trademarks include more processing because registered trademarks need to be unique in their field so as not to impinge on another prior trademark's domain. A patent can be valuable IP, and patents have been correlated with prosperity. Refereed journal articles as a metric are not as highly valued outside of academia as inside, although company executives state in interviews that publications help to attract and keep high-quality staff and also provide additional validation for, and publicity about, their technology.

TABLE G-17 Preferences for Use of Marketing Funds

	Percentage of Responses	
	STTR Awardees	SBIR Awardees
Continue to use the agency's program	37.0	29.6
Use the funding for your own marketing consultant	49.6	47.0
Neither	13.4	23.4
BASE: TOTAL RESPONDENTS ANSWERING QUESTION	119	685

SOURCE: 2011-2014 Survey, Question 52.

TABLE G-18 Number of Patents Related to All Company STTR/SBIR Awards

	Percentage of Responding Companies	
	STTR Awardees	SBIR Awardees
0	47.1	34.7
1 or more	52.9	65.3
1	10.0	14.6
2	14.9	11.4
3	7.2	7.8
4	7.0	5.7
5 to 9	8.2	13.1
10 or more	5.5	12.7
Mean	2.03	5.01
Median	1.00	2.00
BASE: TOTAL COMPANIES ANSWERING QUESTION	155	985

SOURCE: 2011-2014 Survey, Question 12.

The questionnaire also asked questions about intellectual property related to the specific award being surveyed. Forty-two percent of STTR respondents reported receiving at least one patent related to the surveyed technology. Two percent reported receiving five or more related patents. SBIR outcomes were similar (see Table G-19).

Copyrights

About 15 percent of STTR and SBIR respondents reported receiving at least one copyright. Less than 1 percent reported receiving 10 or more.[10]

[10]2011-2014 Survey, Question 38.2.

TABLE G-19 Patents Awarded Related to Surveyed Project

	Percentage of Responses	
	STTR Awardees	SBIR Awardees
0	58.2	55.8
1 or more	41.8	44.2
1	26.0	21.7
2	7.1	11.1
3 or 4	6.6	6.0
5 to 9	1.0	3.6
10 or more	1.0	1.8
Mean	0.8	1.1
Median	0.0	0.0
BASE: TOTAL RESPONDENTS ANSWERING QUESTION	196	1,247

SOURCE: 2011-2014 Survey, Question 38.1.

Trademarks

There was some interest in project-related trademarks, with 28 percent of STTR respondents indicating that at least one had been received.[11] This percentage was slightly higher than that for SBIR respondents (25 percent).

Peer-reviewed Publications

The survey also asked about peer-reviewed publications. Eighty-two percent of STTR respondents indicated that an author at the surveyed company had published at least one scientific paper related to the award. Forty-six percent reported publishing three or more related papers (see Table G-20).

STTR AND COMPANIES

The SBIR/STTR programs have a range of impacts on companies, which affect their ability to work within the innovation ecology of the agency or indeed more generally. Data about companies can help to define the technological space in which the SBIR/STTR programs operate. Finally, a review of the SBIR/STTR share of overall company activities can provide insights into the degree of dependence on SBIR/STTR for individual companies.

[11]2011-2014 Survey, Question 38.3.

Impact on Company Formation

Previous Academy studies have concluded that, for at least some companies, SBIR funding provided opportunities that led directly to company formation. Thirty-six percent of STTR respondents indicated that the program contributed to some degree to company formation (see Table G-21). SBIR companies were somewhat more likely to report that the program supported company formation (41 percent).

TABLE G-20 Peer-Reviewed Scientific Publications Related to the Surveyed Project

	Percentage of Responses	
	STTR Awardees	SBIR Awardees
0	17.7	23.0
1	17.2	19.2
2	19.5	17.5
3 or 4	21.4	19.9
5 to 9	12.6	12.3
10 or more	11.6	8.1
1 or more	82.3	77.0
Mean	4.5	4.4
Median	2.0	2.0
BASE: TOTAL RESPONDENTS ANSWERING QUESTION	215	1,341

SOURCE: 2011-2014 Survey, Question 38.4.

TABLE G-21 Company Founded Because of SBIR-STTR Programs?

	Percentage of Responding Companies		
	STTR Awardees	SBIR Awardees	TOTAL
Yes	12.0	17.1	16.3
In part	24.1	23.8	23.8
No	63.9	59.2	59.8
BASE: TOTAL RESPONDENTS ANSWERING QUESTION	177	1,049	1,226

SOURCE: 2011-2014 Survey, Question 7.

SBIR/STTR Share of Research and Development (R&D) Effort

The survey asked respondents to estimate how much of their company's total R&D effort—defined as man-hours of work for scientists and engineers—was devoted to SBIR/STTR-funded projects in the most recent fiscal year. Overall, 48 percent of STTR respondents indicated that SBIR/STTR funded 10 percent or less of total effort, while 28 percent indicated that it funded more than one-half (see Table G-22).

These data correspond fairly closely to responses from Question 9, which asked what percentage of company revenues during its current year were related to SBIR/STTR awards. Almost one-third of STTR respondents reported zero revenue from SBIR/STTR in the most recent fiscal year, while 24 percent reported receiving more than one-half of all company revenues from SBIR/STTR. Four percent were entirely dependent on SBIR/STTR (see Table G-23).

Prior Use of the SBIR/STTR Programs

Although linear conceptualizations of the innovation process claim that ideas are tested in Phase I, prototyped in Phase II, and commercialized in Phase III, experience shows that real-world development is far more complex. In many cases, multiple iterations are required, or projects must restart with an earlier phase, or multiple efforts are needed to meet specific problems.

The survey asked respondents to indicate how many prior SBIR/STTR Phase I awards were related to the project and technology being surveyed. Table G-24 shows that 20 percent of STTR projects received no other related

TABLE G-22 Percentage of R&D effort Funded by SBIR/STTR

	Percentage of Responding Companies	
	STTR Awardees	SBIR Awardees
0%	28.1	25.4
1-10%	19.4	13.1
11-25%	10.9	14.2
26-50%	13.2	16.9
51-75%	11.8	14.8
76-100%	16.6	15.5
BASE: TOTAL COMPANIES ANSWERING QUESTION	158	1,010

NOTE: Because multiple responses were received from some companies, responses here are weighted to provide the average response per company.
SOURCE: 2011-2014 Survey, Question 10.

SBIR/STTR awards, while 24 percent received at least three additional related awards. These data strongly support the view that innovative products emerge from clusters of activity, rather than simple straight-line development from Phase I to Phase II to commercialization.

TABLE G-23 Percentage of Company Revenues from SBIR/STTR (company's most recent Fiscal Year)

	Percentage of Responding Companies	
	STTR Awardees	SBIR Awardees
0%	31.3	28.7
1-10%	22.2	15.4
11-25%	11.7	14.5
26-50%	10.5	15.7
51-75%	10.9	11.9
76-99%	9.0	11.3
100%	4.4	2.5
BASE: TOTAL RESPONDENTS ANSWERING QUESTION	158	997

SOURCE: 2011-2014 Survey, Question 9.

TABLE G-24 Prior SBIR/STTR Phase I Awards Related to the Surveyed Project

	Percentage of Responses	
	STTR Awardees	SBIR Awardees
0	20.0	20.0
1	37.6	36.8
2	18.8	17.8
3 or 4	11.8	16.3
5 to 9	8.6	6.6
10 or more	3.1	2.5
1 or more	80.0	80.0
Mean	2.1	2.1
Median	1	1
BASE: TOTAL RESPONDENTS ANSWERING QUESTION	255	1,701

SOURCE: 2011-2014 Survey, Question 39.1.

About one-quarter of STTR projects reported no prior related Phase II awards, while 15 percent reported at least three. The median response was one (see Table G-25).

Long-term Impacts on the Recipient Company

SBIR awards have direct effects on specific projects, but they can have a longer-term effect on the trajectory of company development, creating capacity and in some cases providing a critical input that transforms long-term outcomes. The survey asked respondents about this directly. The results are summarized in Table G-26.

These results show an overwhelmingly positive impact. Overall, 79 percent of STTR respondents reported a positive effect, and 32 percent reported a transformative impact. These percentages are somewhat lower than those for SBIR. STTR respondents were also more likely to report a highly negative or disastrous effect (6 percent).

To probe more deeply into this critically important area, respondents were also asked to provide a description of these effects in their own words. Their comments are summarized beginning on the following page, focused on the ways in which SBIR and STTR made a major difference to the company in the long term.

TABLE G-25 Other SBIR or STTR Phase II Awards Related to the Surveyed Project Technology

	Percentage of Responses	
	STTR Awardees	SBIR Awardees
0	22.7	24.8
1	47.0	42.3
2	15.4	17.5
3 or 4	10.5	11.3
5 to 9	3.6	3.1
10 or more	0.8	1.0
1 or more	77.3	75.2
Mean	1.4	1.5
Median	1	1

SOURCE: 2011-2014 Survey, Question 39.2.

TABLE G-26 Long-Term Impacts on Recipient Companies

	Percentage of Responses	
	STTR Awardees	SBIR Awardees
Had a highly positive or transformative effect	32.2	39.2
Had a positive effect	46.4	46.8
Had no effect	15.0	11.2
Had a negative effect	5.2	2.2
Had a highly negative or disastrous effect	1.1	0.6
BASE: TOTAL RESPONDENTS ANSWERING QUESTION	267	1,816

SOURCE: 2011-2014 Survey, Question 57.

Key Aspects of STTR-driven Transformation

It is not easy to summarize the numerous ways in which STTR awards helped to transform recipient companies. What follows is therefore a limited list of impacts drawn from the survey responses. Impacts included the following:

- Supported company formation
- Encouraged faculty to form companies without being forced to leave their academic positions
- Provided first dollars
- Funded areas where venture capital and other funders were not interested
- Supported development of critical company infrastructure
- Opened doors to potential partners
- Helped address niche markets too small for major players/funders
- Funded technology development
- Enabled projects with high levels of technical risk and high potential return
- Supported adaptation of technologies to new uses, markets, and industry sectors
- Funded development of core technology
- Diversified expertise, allowed hiring of specialists
- Gave companies immediate credibility
- Funded researchers to enter business full time
- Transformed company culture to become more market driven
- Created new companies and kept companies in business (that would not exist without STTR funding)

- Helped increase the company's knowledge base applied to later projects
- Expanded the scope and scale of R&D capabilities
- Supported technology development that led to spin-off companies

Overall, the strongest conclusion to be drawn from these responses is that small innovative companies are highly sensitive to the impact of exogenous variables. The sudden withdrawal of a sponsor can crush a company; a single contract can provide funding for 2 or 3 years of growth. Above all, these small companies are highly path dependent: what happens to them at a given moment can dramatically affect long-term outcomes. The butterfly effect could have been invented specifically to apply to these kinds of companies.

In the end, SBIR/STTR can be seen in many cases as a positive exogenous variable: one that provides funding, validation, and often market access not otherwise available. Even though it seems tenuous to link one award to the eventual success of a large corporation, that is, in fact, how some very small companies grow into large ones. The evidence from survey respondents suggests that this positive jolt is not an uncommon effect of these awards.

Industry Sector

Previous analyses of SBIR/STTR have not addressed a potentially important intervening variable: industry sector. It is quite possible that commercialization outcomes may be affected by the average cycle time of product development in different sectors. For example, product cycle time is much shorter in software than in materials or medical devices. Table G-27 shows the distribution of responses by program and sector. For most sectors there are few differences between SBIR and STTR respondents; the former tend to work somewhat more in defense, the latter somewhat more in medical technologies.

COUNTERFACTUALS

It is difficult to tightly determine the impact of a given STTR award. Many factors affect the success and failure of companies and projects, and it is often difficult to determine whether a specific factor was a *necessary* condition for success. Worse still, the large number of factors and the multiple paths to success and failure mean that it is unusual to be able to state with confidence that a particular intervention—in this case an STTR award—constituted a *sufficient* condition for a project's success.

This question was designed to provide an approximate map of activities by sector. There is considerable overlap between some categories, and respondents would have substantial leeway to define sectors differently, so these results should be viewed as highly preliminary.

TABLE G-27 Distribution of Responses by Sector and Phase

Sector	Percentage of Responses STTR Awardees	SBIR Awardees
Aerospace and Defense	39.0	46.5
Aerospace	24.3	29.3
Defense-specific products and services	30.1	36.4
Energy and the Environment	17.6	15.1
Renewable energy production (solar, wind, geothermal, bio-energy, wave)	4.8	3.8
Energy storage and distribution	4.0	3.0
Energy efficiency	5.1	4.0
Other energy or environmental products and services	12.1	9.8
Engineering	36.0	38.9
Engineering services	8.8	9.1
Scientific instruments and measuring equipment	15.1	14.6
Robotics	3.3	3.7
Sensors	14.0	16.7
Other engineering	12.5	15.3
Information Technology	13.6	14.3
Computers and peripheral equipment	2.2	3.0
Telecommunications equipment and services	2.2	2.5
Business and productivity software	1.5	2.6
Data processing and database software and services	5.1	4.4
Media products (including web-, print- and wireless-delivered content)	1.5	1.9
Other IT	3.7	5.6
Materials (including nanotechnology for materials)	16.9	12.9
Medical Technologies	40.4	34.2
Pharmaceuticals	6.6	5.4
Medical devices	12.9	14.5
Biotechnology (including therapeutic, diagnostic, combination)	13.2	11.4
Health IT (including mobile, big data, training modules)	2.9	2.0
Research tools	10.3	8.0
Education materials	3.7	2.4
Other medical products and services	4.4	3.9
Other (please specify)	8.1	11.6
BASE: TOTAL RESPONDENTS ANSWERING QUESTION	272	1,861

NOTE: Answers do not sum to 100 percent because respondent could select more than one sector.
SOURCE: 2011-2014 Survey, Question 20.

Project Go-ahead Absent STTR Funding

One approach has been to ask recipients for their own views on the impact of the program on their project or company. In particular, the survey asked whether the project would have been undertaken absent STTR funding and whether the scope and timing would have been affected. Responses are summarized in Table G-28.

Only about 9 percent of the STTR respondents indicated that there was even a probability that the project would have proceeded without program funding. In contrast, 76 percent thought the project would probably not have proceeded absent STTR funding; 38 percent were definite that the project would not have proceeded.

Project Scope Absent STTR Funding

A second area of review concerns the impact of funding on the project's scope. It seems likely that additional funding in the form of STTR money would lead to an expansion of project scope. The analysis in this case focused only on the respondents who were certain that the project would have proceeded absent program funding. Results are summarized in Table G-29.

Although most respondents indicated that the absence of program funding would have limited the scope of the project, some respondents indicated that the ambitions of the project were limited by participation in the program, most likely to address the specific requirements of SBIR/STTR awards.

Project Delays Absent STTR Funding

As with project scope, the immediate supposition is that, absent STTR funding, projects would have been delayed while other funding was identified and acquired. However, as we will see when considering program operations

TABLE G-28 Project Undertaken in the Absence of This STTR Award

	Percentage of Responses	
	STTR Awardees	SBIR Awardees
Definitely yes	1.8	1.9
Probably yes	7.3	7.6
Uncertain	15.3	15.4
Probably not	37.2	40.3
Definitely not	38.3	34.8
BASE: TOTAL RESPONDENTS ANSWERING QUESTION	274	1,867

SOURCE: 2011-2014 Survey, Question 24.

TABLE G-29 STTR Impact on Project Scope

	Percentage of Responses		
	STTR Awardees	SBIR Awardees	TOTAL
Broader in scope	8.0	8.6	8.5
Similar in scope	16.0	28.0	26.5
Narrower in scope	76.0	63.4	65.0
BASE: COMPANY WOULD HAVE UNDERTAKEN PROJECT IN THE ABSENCE OF THE AWARD	25	175	200

SOURCE: 2011-2014 Survey, Question 25.

later in this report, STTR awards carry delays of their own, which can in some cases be substantial. Thus this survey question seeks to determine a balance between delays imposed by the need for new funding and delays inherent in the STTR program.

Ninety-six percent of the 25 STTR respondents who were sure that the project would have proceeded absent STTR funding agreed that the absence of STTR funding would have delayed the project by at least 3 months (Table G-30). Twenty-four percent projected a delay of at least 12 months. Given that gaps and delays can have a significant impact on small companies with few other resources for retaining their technical teams, this is potentially an important result. SBIR projects reported being somewhat less susceptible to delays caused by the absence of Phase II funding.

Data for STTR are similar to those for SBIR, reinforcing the limited participation of both woman- and minority-owned firms in the programs. Firms owned by minorities other than Asian-Americans were especially poorly represented among respondents from both STTR and SBIR projects.

TABLE G-30 Likely Delay Absent STTR Funding

	Percentage of Responses	
	STTR Awardees	SBIR Awardees
Less than 3 months	4.0	11.5
3 to 6 months	24.0	14.5
7 to 12 months	48.0	36.4
Over 12 months	24.0	37.6
Average months	15.6	15.0
Median months	12	12
BASE: COMPANY WOULD HAVE UNDERTAKEN PROJECT IN THE ABSENCE OF THE AWARD	25	165

SOURCE: 2011-2014 Survey, Question 26A.

WOMEN AND MINORITIES

Although the participation of women and disadvantaged groups is not a formal objective of the STTR program, the 2011-2014 Survey did explore levels of participation both for companies and for PIs. Results are provided in tables G-31 and G-32.

TABLE G-31 Participation of Woman- and Minority-owned Companies in the STTR/SBIR Program

	Percentage of Responses	
	STTR Awardees	SBIR Awardees
Woman-owned	8.4	10.6
Minority-owned	9.1	9.1
Asian-Indian	4.6	3.5
Asian-Pacific	2.3	3.4
Black	1.1	0.3
Hispanic	1.1	1.5
Native American	0.4	0.2
Other	0.4	0.3
Not woman- nor minority-owned	83.7	82.3
BASE: TOTAL RESPONDENTS ANSWERING QUESTION	263	1,789

SOURCE: 2011-2014 Survey, Question 15.

TABLE G-32 Participation of Female and Minority PIs in the STTR/SBIR Program

	Percentage of Responses	
	STTR Awardees	SBIR Awardees
Woman	10.0	8.9
Minority	14.1	11.3
Asian-Indian	4.1	4.1
Asian-Pacific	4.1	5.1
Black	0.7	0.2
Hispanic	3.0	1.3
Native American	0.7	0.2
Other	1.5	0.5
Not a woman nor a minority	79.3	81.2
BASE: TOTAL RESPONDENTS ANSWERING QUESTION	270	1,856

SOURCE: 2011-2014 Survey, Question 16.

Data for PIs are similar, although both female and minority participation are greater than for woman-and minority-owned firms. Again, minorities other than Asian-Americans were especially poorly represented.

Appendix H

Glossary

ARRA	American Recovery and Reinvestment Act of 2009
CCT	Center Chief Technologist
CRADA	Cooperative Research and Development Agreement
DoD	Department of Defense
DoE	Department of Energy
IC	Institutes or Centers
I/UCRC	Industry and University Cooperative Research Center
NASA	National Aeronautics and Space Administration
NIH	National Institutes of Health
NSF	National Science Foundation
PI	Principal Investigator
RI	Research Institution
SBA	Small Business Administration
TPOC	Technical Point of Contact
TRL	Technology Readiness Level
TTO	Technology Transfer Office

Appendix I

Agenda: Workshop on the Small Business Technology Transfer Program

May 1, 2015

Lecture Room
National Academy of Sciences
2100 C Street, NW
Washington, DC

9:00 AM **Introduction**
Jacques Gansler, University of Maryland

9:10 AM **Panel I: Agency Perspectives on STTR**
Moderator: Sujai Shivakumar, The National Academies

Christopher Rinaldi, Department of Defense
Matthew Portnoy, National Institutes of Health
Kurt Marek, National Heart, Lung, and Blood Institute
Manny Oliver, Department of Energy
Barry Johnson, National Science Foundation
Joseph Grant, National Aeronautics and Space Agency

10:30 AM **Coffee Break**

10:45 AM **Panel II: Small Business Perspectives on STTR**
Moderator: Michael Borrus, XSeed Capital

Anthony Mulligan, Hydronalix, Inc.
Jay Rozzi, Creare LLC
Terry Grimm, Niowave, Inc.

12:00 PM	**Panel III: Policy Roundtable** *Moderator: Patrick Windham, Technology Policy International* *Jere Glover, Small Business Technology Council* *David Day, University of Florida* *James Woodell, APLU* *Kevin Wheeler, Senate Committee on Small Business and Entrepreneurship*
12:35 PM	**Closing Remarks** *Jacques Gansler, University of Maryland*
12:45 PM	**Adjourn**

Appendix J
Bibliography

Acs, Z., and D. Audretsch. 1988. "Innovation in Large and Small Firms: An Empirical Analysis." *The American Economic Review* 78(4):678-690.

Acs, Z., and D. Audretsch. 1990. *Innovation and Small Firms.* Cambridge, MA: MIT Press.

Adelstein, F. 2006. Live Forensics: Diagnosing Your System Without Killing It First, http://frank.notfrank.com/Papers/CACM06.pdf. Accessed July 17, 2014.

Advanced Technology Program. 2001. *Performance of 50 Completed ATP Projects, Status Report 2.* National Institute of Standards and Technology Special Publication 950-2. Washington, DC: Advanced Technology Program/National Institute of Standards and Technology/U.S. Department of Commerce.

Alic, J. 1987. "Evaluating Competitiveness at the Office of Technology Assessment," *Technology in Society* 9(1):1-17.

Alic, J. A., L. Branscomb, H. Brooks, A. B. Carter, and G. L. Epstein. 1992. *Beyond Spinoff: Military and Commercial Technologies in a Changing World.* Boston, MA: Harvard Business School Press.

American Association for the Advancement of Science. R&D Funding Update on NSF in the FY2007.

American Psychological Association. 2002. "Criteria for Evaluating Treatment Guidelines." *American Psychologist* 57(12):1052-1059.

Archibald, R., and D. Finifter. 2003. "Evaluating the NASA Small Business Innovation Research Program: Preliminary Evidence of a Tradeoff Between Commercialization and Basic Research." *Research Policy* 32:605-619.

Archibugi, D., A. Filippetti, and M. Frenz. 2013. "Economic Crisis and Innovation: Is Destruction Prevailing Over Accumulation?" *Research Policy* 42(2):303-314.

Arrow, K. 1962. "Economic welfare and the allocation of resources for invention." Pp. 609-625 in *The Rate and Direction of Inventive Activity: Economic and Social Factors*. Princeton, NJ: Princeton University Press.

Arrow, K. 1973. "The theory of discrimination." Pp. 3-31 in *Discrimination in Labor Market*. Orley Ashenfelter and Albert Rees, eds. Princeton, NJ: Princeton University Press.

Audretsch, D. B. 1995. *Innovation and Industry Evolution.* Cambridge, MA: MIT Press.

Audretsch, D. B., and M. P. Feldman. 1996. "R&D Spillovers and the Geography of Innovation and Production." *American Economic Review* 86(3):630-640.

Audretsch, D. B., and P. E. Stephan. 1996. "Company-Scientist Locational Links: The Case of Biotechnology." *American Economic Review* 86(3):641-642.

Audretsch, D., and R. Thurik. 1999. *Innovation, Industry Evolution, and Employment.* Cambridge, MA: MIT Press.

Azoulay, P., T. Stuart, and Y. Wang. 2014. "Matthew: Effect or Fable?" *Management Science* 60(1):92-109.

Baker, A. No date. "Commercialization Support at NSF." Draft.

Baker, J. A. K., and K. J. Thurber. 2011. *Developing Computer Systems Requirements.* Ithaca, NY: Digital Systems Press.

Barfield, C., and W. Schambra, eds. 1986. *The Politics of Industrial Policy.* Washington, DC: American Enterprise Institute for Public Policy Research.

Baron, J. 1998. "DoD SBIR/STTR Program Manager." Comments at the Methodology Workshop on the Assessment of Current SBIR Program Initiatives, Washington, DC, October.

Barry, C. B. 1994. "New Directions in Research on Venture Capital Finance." *Financial Management* 23 (Autumn):3-15.

Bator, F. 1958. "The Anatomy of Market Failure." *Quarterly Journal of Economics* 72:351-379.

Berberich, S. 2006. "$29.3M Gates Grant Boosts Sanaria." *Washington Business Journal* December 15.

Biemer, P. P., and L. E. Lyberg. 2003. *Introduction to Survey Quality*. New York: John Wiley & Sons.

Bingham, R. 1998. *Industrial Policy American Style: From Hamilton to HDTV.* New York: M.E. Sharpe.

Birch, D. 1981. "Who Creates Jobs." *The Public Interest* 65 (Fall):3-14.

Bouchie, A. 2003. "Increasing Number of Companies Found Ineligible for SBIR Funding,"
Nature Biotechnology 21(10):1121-1122.

Branscomb, L. M., K. P. Morse, M. J. Roberts, and D. Boville. 2000. *Managing Technical Risk: Understanding Private Sector Decision-Making on Early Stage Technology Based Projects.* Washington, DC: Department of Commerce/National Institute of Standards and Technology.

Branscomb, L. M., and P. E. Auerswald. 2001. *Taking Technical Risks: How Innovators, Managers, and Investors Manage Risk in High-Tech Innovations*, Cambridge, MA: MIT Press.

Branscomb, L. M., and P. E. Auerswald. 2002. *Between Invention and Innovation: An Analysis of Funding for Early-Stage Technology Development.* Gaithersburg, MD: National Institute of Standards and Technology.

Branscomb, L. M., and P. E. Auerswald. 2003. "Valleys of Death and Darwinian Seas: Financing the Invention to Innovation Transition in the United States." *The Journal of Technology Transfer* 28(3-4).

Branscomb, L. M., and J. Keller. 1998. *Investing in Innovation: Creating a Research and Innovation Policy.* Cambridge, MA: MIT Press.

Brav, A., and P. A. Gompers. 1997. "Myth or Reality?: Long-Run Underperformance of Initial Public Offerings; Evidence from Venture Capital and Nonventure Capital-Backed IPOs." *Journal of Finance* 52:1791-1821.

Brodd, R. J. 2005. "Factors Affecting U.S. Production Decisions: Why Are There No Volume Lithium-Ion Battery Manufacturers in the United States?" ATP Working Paper No. 05-01, June.

Brown, G., and J. Turner. 1999. "Reworking the Federal Role in Small Business Research." *Issues in Science and Technology* XV(4 Summer):51-58.

Bush, V. 1946. *Science—the Endless Frontier.* Republished in 1960 by U.S. National Science Foundation, Washington, DC.

Cahill, P. 2000. "Fast track: Is it speeding commercialization of Department of Defense Small Business Innovation Research Projects?" In National Research Council, *The Small Business Innovation Research Program: An Assessment of the Department of Defense Fast Track Initiative.* Washington, DC: National Academy Press.

Carden, S. D., and O. Darragh. 2004. "A Halo for Angel Investors." *The McKinsey Quarterly* 1.

Caves, R.E. 1998. "Industrial Organization and New Findings on the Turnover and Mobility of Firms." *Journal of Economic Literature* 36(4):1947-1982.

Ceulemans, S., and J. K. Kolls. 2013. "Can the SBIR and STTR Programs Advance Research Goals?" *Nature Immunology* 14(3):192-195.

Christensen, C. 1997. *The Innovator's Dilemma.* Boston, MA: Harvard Business School Press.

Christensen, C., and M. Raynor. 2003. *Innovator's Solution*, Boston, MA: Harvard Business School.

Clinton, W. J. 1994. *Economic Report of the President.* Washington, DC: U.S. Government Printing Office.

Clinton, W. J. 1994. *The State of Small Business.* Washington, DC: U.S. Government Printing Office.

Coburn, C., and D. Bergland. 1995. *Partnerships: A Compendium of State and Federal Cooperative Technology Programs.* Columbus, OH: Battelle.

Cochrane, J. H. 2005. "The Risk and Return of Venture Capital." *Journal of Financial Economics* 75(1):3-52.

Cohen, L. R., and R. G. Noll. 1991. *The Technology Pork Barrel.* Washington, DC: The Brookings Institution.

Congressional Commission on the Advancement of Women and Minorities in Science, Engineering, and Technology Development. 2000. *Land of Plenty: Diversity as America's Competitive Edge in Science, Engineering and Technology.* Washington, DC: National Science Foundation/U.S. Government Printing Office.

Cooper, R. G. 2001. Winning at New Products: Accelerating the Process from Idea to Launch. In Dawnbreaker, Inc. 2005. "The Phase III Challenge: Commercialization Assistance Programs 1990-2005." White paper. July 15.

Cooper, R. S. 2003. "Purpose and Performance of the Small Business Innovation Research (SBIR) Program." *Small Business Economics* 20(2):137-151.

Council of Economic Advisers. 1995. *Supporting Research and Development to Promote Economic Growth: The Federal Government's Role.* Washington, DC: Council of Economic Advisors.

Council on Competitiveness. 2005. *Innovate America: Thriving in a World of Challenge and Change.* Washington, DC: Council on Competitiveness.

Crane, G., and J. Sohl. 2004. "Imperatives for Venture Success: Entrepreneurs Speak." *The International Journal of Entrepreneurship and Innovation* May. Pp. 99-106.

Cutler, D. 2005. *Your Money or Your Life.* New York: Oxford University Press.

Cycyota, C. S., and D. A. Harrison. 2006. "What (Not) to Expect When Surveying Executives: A Meta-Analysis of Top Manager Response." *Organizational Research Methods* 9:133-160.

Czarnitzki, D., and A. Fier. 2002. "Do Innovation Subsidies Crowd out Private Investment? Evidence from the German Service Sector." ZEW Discussion Papers, No. 02-04.

Dalton, A. B., S. Collins, E. Muñoz, J. Razal1, V. H. Ebron, J. Ferraris, J. Coleman, B. Kim, and R. Baughman. 2003. "Super-Tough Carbon-Nanotube Fibres." *Nature* 423(4):703.

David, P. A., B. H. Hall, and A. A. Tool. 1999. "Is Public R&D a Complement or Substitute for Private R&D? A Review of the Econometric Evidence." NBER Working Paper 7373. October.

Davidsson, P. 1996. "Methodological Concerns in the Estimation of Job Creation in Different Firm Size Classes." Working Paper. Jönköping International Business School.

Davis, S. J., J. Haltiwanger, and S. Schuh. 1994. "Small Business and Job Creation: Dissecting the Myth and Reassessing the Facts." *Business Economics* 29(3):113-122.

Dawnbreaker, Inc. 2005. "The Phase III Challenge: Commercialization Assistance Programs 1990-2005." White paper. July 15.

Dertouzos, M. L. 1989. *Made in America: The MIT Commission on Industrial Productivity.* Cambridge, MA: MIT Press.

Dertouzos, M. L., R. Lester, and R. Solow. 1989. *Made in America: The MIT Commission on Industrial Productivity*. Cambridge, MA: The MIT Press.

Dess, G. G., and D. W. Beard. 1984. "Dimensions of Organizational Task Environments." *Administrative Science Quarterly* 29:52-73.

Devenow, A., and I. Welch. 1996. "Rational Herding in Financial Economics." *European Economic Review* 40(April):603-615.

Dillman, D. 2000. *Mail and Internet Surveys: The Tailored Design Method.* 2nd Edition. Toronto, Ontario: John Wiley and Sons, Inc.

DoE Opportunity Forum. 2005. "Partnering and Investment Opportunities for the Future." Tysons Corner, VA. October 24-25.

Eckstein, O. 1984. *DRI Report on U.S. Manufacturing Industries.* New York: McGraw Hill.

Eisinger, P. K. 1988. *The Rise of the Entrepreneurial State: State and Local Economic Development Policy in the United States.* Madison, WI: University of Wisconsin Press.

Ernst and Young. 2007. "U.S. Venture Capital Investment Increases to 8 percent to $6.96 Billion in First Quarter of 2007." April 23.

Evenson, R., P. Waggoner, and P. Ruttan. 1979. "Economic Benefits from Research: An Example from Agriculture." *Science* 205(14 September):1101-1107.

Feenstra, D. 2014. "Public Support of Innovation in Entrepreneurial Firms." *Journal of Applied Management and Entrepreneurship* 19(2):135.

Feldman, M. P. 1994. *The Geography of Knowledge.* Boston, MA: Kluwer Academic.

Feldman, M. P. 1994. "Knowledge Complementarity and Innovation." *Small Business Economics* 6(5):363-372.

Feldman, M. P. 2001. "Assessing the ATP: Halo Effects and Added Value." In National Research Council, *The Advanced Technology Program: Assessing Outcomes*. Washington, DC: National Academy Press.

Feldman, M. P., and M. R. Kelley. 2001. *Winning an Award from the Advanced Technology Program: Pursuing R&D Strategies in the Public Interest and Benefiting from a Halo Effect.* NISTIR 6577. Washington, DC: Advanced Technology Program/National Institute of Standards and Technology/U.S. Department of Commerce.

Fenn, G. W., N. Liang, and S. Prowse. 1995. *The Economics of the Private Equity Market.* Washington, DC: Board of Governors of the Federal Reserve System.

Financial Times. 2004. "Qinetiq set to make its first US acquisition," September 8.

Fischer, E., and A. R. Reuber. 2003. "Support for Rapid-Growth Firms: A Comparison of the Views of Founders, Government Policymakers, and Private Sector Resource Providers," *Journal of Small Business Management* 41(4):346-365.

Flamm, K. 1988. *Creating the Computer.* Washington, DC: The Brookings Institution.

Flender, J. O., and R. S. Morse. 1975. *The Role of New Technical Enterprise in the U.S. Economy.* Cambridge, MA: MIT Development Foundation.

Freear, J., and W. E. Wetzel Jr. 1990. "Who Bankrolls High-Tech Entrepreneurs?" *Journal of Business Venturing* 5:77-89.
Freeman, C., and L. Soete. 1997. *The Economics of Industrial Innovation.* Cambridge, MA: MIT Press.
Galbraith, J. K. 1957. *The New Industrial State.* Boston: Houghton Mifflin.
Gallagher, S. 2012. "Here Come the Inflate-a-Bots: iRobot's AIR Blow Up Bot Prototypes." ARS Technica.
Galope, R. V. 2014. "What Types of Start-ups Receive Funding from the Small Business Innovation Research (SBIR) Program?: Evidence from the Kauffman Firm Survey." *Journal of Technology Management & Innovation* 9(2):17-28.
Geroski, P. A. 1995. "What Do We Know About Entry?" *International Journal of Industrial Organization* 13(4):421-440.
Geshwiler, J., J. May, and M. Hudson. 2006. *State of Angel Groups*. Kansas City, MO: Kauffman Foundation.
Gicheva, D., and A. N. Link. 2015. "The Gender Gap in Federal and Private Support for Entrepreneurship." *Small Business Economics* (2015):1-5.
Gompers, P. A., and J. Lerner. 1977. "Risk and Reward in Private Equity Investments: The Challenge of Performance Assessment." *Journal of Private Equity* 1:5-12.
Gompers, P. A. 1995. "Optimal Investment, Monitoring, and the Staging of Venture Capital." *Journal of Finance* 50:1461-1489.
Gompers, P. A., and J. Lerner. 1996. "The Use of Covenants: An Empirical Analysis of Venture Partnership Agreements." *Journal of Law and Economics* 39:463-498.
Gompers, P. A., and J. Lerner. 1998. "Capital Formation and Investment in Venture Markets: A Report to the NBER and the Advanced Technology Program." Unpublished working paper. Harvard University.
Gompers, P. A., and J. Lerner. 1998. "What Drives Venture Capital Fund-Raising?" Unpublished working paper. Harvard University.
Gompers, P. A., and J. Lerner. 1999. "An Analysis of Compensation in the U.S. Venture Capital Partnership." *Journal of Financial Economics* 51(1):3-7.
Gompers, P. A., and J. Lerner. 1999. *The Venture Cycle.* Cambridge, MA: MIT Press.
Good, M. L. 1995. Prepared testimony before the Senate Commerce, Science, and Transportation Committee, Subcommittee on Science, Technology, and Space (photocopy, U.S. Department of Commerce).
Goodnight, J. 2003. Presentation at National Research Council Symposium. "The Small Business Innovation Research Program: Identifying Best Practice." Washington, DC, May 28.
Graham, O. L. 1992. *Losing Time: The Industrial Policy Debate.* Cambridge, MA: Harvard University Press.
Greenwald, B. C., J. E. Stiglitz, and A. Weiss. 1984. "Information Imperfections in the Capital Market and Macroeconomic Fluctuations." *American Economic Review Papers and Proceedings* 74:194-199.

Griliches, Z. 1990. *The Search for R&D Spillovers.* Cambridge, MA: Harvard University Press.

Groves, R. M., D. A. Dillman, J. L. Eltinge, and R. J. A. Little, eds. 2002. *Survey Nonresponse*. New York: Wiley.

Groves, R. M., F. J. Fowler, Jr., M. P. Couper, J. M. Lepkowski, E. Singer, and R. Tourangeau. 2004. *Survey Methodology.* Hoboken, NJ: John Wiley & Sons, Inc.

Hall, B. H. 1992. "Investment and Research and Development: Does the Source of Financing Matter?" Working Paper No. 92-194, Department of Economics/University of California at Berkeley.

Haltiwanger, J., and C. J. Krizan. 1999. "Small Businesses and Job Creation in the United States: The Role of New and Young Businesses" in *Are Small Firms Important? Their Role and Impact*, Zoltan J. Acs, ed. Dordrecht: Kluwer.

Hamberg, D. 1963. "Invention in the Industrial Research Laboratory." *Journal of Political Economy* (April):95-115.

Hao, K. Y., and A. B. Jaffe. 1993. "Effect of Liquidity on Firms' R&D Spending." *Economics of Innovation and New Technology* 2:275-282.

Hebert, R. F., and A. N. Link. 1989. "In Search of the Meaning of Entrepreneurship." *Small Business Economics* 1(1):39-49.

Held, B., T. Edison, S. L. Pfleeger, P. Anton, and J. Clancy. 2006. *Evaluation and Recommendations for Improvement of the Department of Defense Small Business Innovation Research (SBIR) Program.* Arlington, VA: RAND National Defense Research Institute.

Henrekson, M., and D. Johansson. 2009. "Competencies and Institutions Fostering High-Growth Firms." *Foundations and Trends in Entrepreneurship* 5(1):1-80.

Hess, J. 2014. "Clinical Operations: Accelerating Trials, Allocating Resources and Measuring Performance." *Cutting Edge Information.* October 12.

Himmelberg, C. P., and B. C. Petersen. 1994. "R&D and Internal Finance: A Panel Study of Small Firms in High-Tech Industries." *Review of Economics and Statistics* 76:38-51.

Hong, S., S. Myung. 2007. "Nanotube Electronics: A Flexible Approach to Obility. *Nature Nanotechnology* 2(4):207-208.

Howell, S. 2015. "DOE SBIR Evaluation: Impact of Small Grants on Subsequent Venture Capital Investment, Patenting, and Achieving Revenue." Paper presented at the National Academy of Sciences, Engineering, and Medicine Workshop on the Economics of Entrepreneurship. June 29.

Hubbard, R. G. 1998. "Capital-Market Imperfections and Investment." *Journal of Economic Literature* 36:193-225.

Huntsman, B., and J. P. Hoban Jr. 1980. "Investment in New Enterprise: Some Empirical Observations on Risk, Return, and Market Structure." *Financial Management* 9(Summer):44-51.

Institute of Medicine. 1998. "The Urgent Need to Improve Health Care Quality." National Roundtable on Health Care Quality. *Journal of the American Medical Association* 280(11):1003, September 16.

Jacobs, T. 2002. "Biotech Follows Dot.com Boom and Bust." *Nature* 20(10):973.

Jaffe, A. B. 1996. "Economic Analysis of Research Spillovers: Implications for the Advanced Technology Program." Washington, DC: Advanced Technology Program/National Institute of Standards and Technology/U.S. Department of Commerce).

Jaffe, A. B. 1998. "The Importance of 'Spillovers' in the Policy Mission of the Advanced Technology Program." *Journal of Technology Transfer* (Summer).

Jarboe, K. P., and R. D. Atkinson. 1998. "The Case for Technology in the Knowledge Economy: R&D, Economic Growth and the Role of Government." Washington, DC: Progressive Policy Institute.

Jewkes, J., D. Sawers, and R. Stillerman. 1958. *The Sources of Invention.* New York: St. Martin's Press.

Johnson, W. 2004. "Delivering Combat Power to the Fleet." *Naval Engineers Journal*, Fall 2004, pp. 3-5.

Johnson, T., and L. Owens. 2003. "Survey Response Rate Reporting in the Professional Literature." Paper presented at the 58th Annual Meeting of the American Association for Public Opinion Research. Nashville, TN. May.

Kaplowitz, M. D., T. D. Hadlock, and R. Levine. 2004. "A Comparison of Web and Mail Survey Response Rates." *Public Opinion Quarterly* 68(1):94-101.

Kauffman Foundation. 2014. About the Foundation. http://www.kauffman.org/foundation.cfm.

Kleinman, D. L. 1995. *Politics on the Endless Frontier: Postwar Research Policy in the United States.* Durham, NC: Duke University Press.

Kolosnjaj, J., H. Szwarc, and F. Moussa. 2007. "Toxicity Studies of Carbon Nanotubes." *Advances in Experimental Medicine and Biology* 620:181-204.

Kortum, S., and J. Lerner. 1998. "Does Venture Capital Spur Innovation?" NBER Working Paper No. 6846, National Bureau of Economic Research.

Krugman, P. 1990. *Rethinking International Trade.* Cambridge, MA: MIT Press.

Krugman, P. 1991. *Geography and Trade.* Cambridge, MA: MIT Press.

Kurtz, A. 2015. "The NCI SBIR Program: An Overview of New Funding Opportunities and Strategies for Employing Lean Startup Tools to Drive Success in Your Small Business." AACR presentation. April 20.

Lanahan, L. 2015. "Multilevel Public Funding for Small Business Innovation: A Review of US State SBIR Match Programs." *The Journal of Technology Transfer* (2015):1-30.

Lanahan, L., and M. P. Feldman. 2015. "Multilevel Innovation Policy Mix: A Closer Look at State Policies that Augment the Federal SBIR Program." *Research Policy* 44(7):1387-1402.

Langlois, R. N., and P. L. Robertson. 1996. "Stop Crying over Spilt Knowledge: A Critical Look at the Theory of Spillovers and Technical Change." Paper prepared for the MERIT Conference on Innovation, Evolution, and Technology. Maastricht, Netherlands, August 25-27.

Langlois, R. N. 2001. "Knowledge, Consumption, and Endogenous Growth." *Journal of Evolutionary Economics* 11:77-93.

Lebow, I. 1995. *Information Highways and Byways: From the Telegraph to the 21st Century.* New York: Institute of Electrical and Electronic Engineering.

Ledford, H. 2014. "Indirect costs: Keeping the lights on." *Nature* 515(7527): 326-9.

Lerner, J. 1994. "The Syndication of Venture Capital Investments." *Financial Management* 23-(Autumn):16-27.

Lerner, J. 1995. "Venture Capital and the Oversight of Private Firms." *Journal of Finance* 50:301-318.

Lerner, J. 1996. "The Government as Venture Capitalist: The Long-Run Effects of the SBIR Program." Working Paper No. 5753, National Bureau of Economic Research.

Lerner, J. 1998. "Angel Financing and Public Policy: An Overview." *Journal of Banking and Finance* 22(6-8):773-784.

Lerner, J. 1999. "The Government as Venture Capitalist: The Long-Run Effects of the SBIR Program." *Journal of Business* 72(3):285-297.

Lerner, J. 1999. "Public Venture Capital: Rationales and Evaluation." In *The SBIR Program: Challenges and Opportunities.* Washington, DC: National Academy Press.

Levy, D. M., and N. Terleckyk. 1983. "Effects of Government R&D on Private R&D Investment and Productivity: A Macroeconomic Analysis." *Bell Journal of Economics* 14:551-561.

Liles, P. 1977. *Sustaining the Venture Capital Firm.* Cambridge, MA: Management Analysis Center.

Link, A. N. 1998. "Public/Private Partnerships as a Tool to Support Industrial R&D: Experiences in the United States." Paper prepared for the working group on Innovation and Technology Policy of the OECD Committee for Science and Technology Policy, Paris.

Link, A. N., and J. Rees. 1990. "Firm Size, University Based Research and the Returns to R&D." *Small Business Economics* 2(1):25-32.

Link, A. N., and J. T. Scott. 1998. "Assessing the Infrastructural Needs of a Technology-Based Service Sector: A New Approach to Technology Policy Planning." *STI Review* 22:171-207.

Link, A. N., and J. T. Scott. 1998. *Overcoming Market Failure: A Case Study of the ATP Focused Program on Technologies for the Integration of Manufacturing Applications (TIMA).* Draft final report submitted to the Advanced Technology Program. Gaithersburg, MD: National Institute of Technology. October.

Link, A. N., and J. T. Scott. 1998. *Public Accountability: Evaluating Technology-Based Institutions.* Norwell, MA: Kluwer Academic.

Link, A. N., and J. T. Scott. 2005. *Evaluating Public Research Institutions: The U.S. Advanced Technology Program's Intramural Research Initiative.* London: Routledge.

Link, A. N., and J. T. Scott. 2012. "The Exploitation of Publicly Funded Technology." *The Journal of Technology Transfer* 37(3):375-383.

Link, A. N., and J. T. Scott. 2012. "The Small Business Innovation Research Program." *Issues in Science and Technology* 28(4):89-92.

Longini, P. 2003. "Hot Buttons for NSF SBIR Research Funds." Pittsburgh Technology Council. *TechyVent*. November 27.

Malone, T. 1995. *The Microprocessor: A Biography.* Hamburg, Germany: Springer Verlag/Telos.

Mankins, J. C. 1995. *Technology Readiness Levels: A White Paper.* Washington, DC: NASA Office of Space Access and Technology. Advanced Concepts Office.

Mann, D., Q. Wang, K. Goodson, and H. Dai. 2005. "Thermal Conductance of an Individual Single-Wall Carbon Nanotube Above Room Temperature." *Nano Letters* 6(1):96-100.

Mansfield, E. 1985. "How Fast Does New Industrial Technology Leak Out?" *Journal of Industrial Economics* 34(2).

Mansfield, E. 1996. *Estimating Social and Private Returns from Innovations Based on the Advanced Technology Program: Problems and Opportunities.* Unpublished report.

Mansfield, E., J. Rapoport, A. Romeo, S. Wagner, and G. Beardsley. 1977. "Social and Private Rates of Return from Industrial Innovations." *Quarterly Journal of Economics* 91:221-240.

Martin, Justin. 2002. "David Birch." *Fortune Small Business* (December 1).

Mathers, E. 2015. "Life Science Startups Looking for New Sources of Funding." *Scale Finance* July 14.

McCraw, T. 1986. "Mercantilism and the Market: Antecedents of American Industrial Policy." In C. Barfield and W. Schambra, eds. *The Politics of Industrial Policy.* Washington, DC: American Enterprise Institute for Public Policy Research.

Mervis, J. D. 1996. "A $1 Billion 'Tax' on R&D Funds." *Science* 272:942–944.

Morgenthaler, D. 2000. "Assessing Technical Risk," in *Managing Technical Risk: Understanding Private Sector Decision Making on Early Stage Technology-Based Project.* L. M. Branscomb, K. P. Morse, and M. J. Roberts, eds. Gaithersburg, MD: National Institute of Standards and Technology.

Mowery, D. 1998. "Collaborative R&D: How Effective Is It?" *Issues in Science and Technology* (Fall):37-44.

Mowery, D., and N. Rosenberg. 1989. *Technology and the Pursuit of Economic Growth.* New York: Cambridge University Press.

Mowery, D., and N. Rosenberg. 1998. *Paths of Innovation: Technological Change in 20th Century America.* New York: Cambridge University Press.

Mowery, D. C., et al. 2001. "The Growth of Patenting and Licensing by US Universities: An Assessment of the Effects of the Bayh–Dole Act of 1980." *Research Policy* 30(1):99-119.

Mowery, D. C. 1999. "America's Industrial Resurgence (?): An Overview." in National Research Council. *U.S. Industry in 2000: Studies in Competitive Performance*. D.C. Mowery, ed. Washington, DC: National Academy Press, p. 1.

Murphy, L. M., and P. L. Edwards. 2003. *Bridging the Valley of Death—Transitioning from Public to Private Sector Financing*. Golden, CO: National Renewable Energy Laboratory. May.

Myers, S., R. L. Stern, and M. L. Rorke. 1983. *A Study of the Small Business Innovation Research Program*. Lake Forest, IL: Mohawk Research Corporation.

Myers, S. C., and N. Majluf. 1984. "Corporate Financing and Investment Decisions When Firms Have Information That Investors Do Not Have." *Journal of Financial Economics* 13:187-221.

National Academies of Sciences, Engineering, and Medicine. 2015. *Innovation, Diversity, and the SBIR/STTR Programs*. Washington, DC: The National Academies Press.

National Academies of Sciences, Engineering, and Medicine. 2015. *SBIR/STTR at the National Institutes of Health*. Washington, DC: The National Academies Press.

National Academies of Sciences, Engineering, and Medicine. 2015. *SBIR at the National Science Foundation*. Washington, DC: The National Academies Press.

National Aeronautics and Space Administration. 2002. "Small Business/SBIR: NICMOS Cryocooler—Reactivating a Hubble Instrument." *Aerospace Technology Innovation* 10(4):19-21.

National Aeronautics and Space Administration. 2005. "The NASA SBIR and STTR Programs Participation Guide." http://sbir.gsfc.nasa.gov/SBIR/zips/guide.pdf.

National Institutes of Health. 2003. "Road Map for Medical Research." http://nihroadmap.nih.gov/.

National Research Council. 1986. *The Positive Sum Strategy: Harnessing Technology for Economic Growth*. Washington, DC: National Academy Press.

National Research Council. 1987. *Semiconductor Industry and the National Laboratories: Part of a National Strategy*. Washington, DC: National Academy Press.

National Research Council. 1991. *Mathematical Sciences, Technology, and Economic Competitiveness*. Washington, DC: National Academy Press.

National Research Council. 1992. *The Government Role in Civilian Technology: Building a New Alliance*. Washington, DC: National Academy Press.

National Research Council. 1995. *Allocating Federal Funds for R&D*. Washington, DC: National Academy Press.

National Research Council. 1996. *Conflict and Cooperation in National Competition for High-Technology Industry*. Washington, DC: National Academy Press.
National Research Council. 1997. *Review of the Research Program of the Partnership for a New Generation of Vehicles: Third Report*. Washington, DC: National Academy Press.
National Research Council. 1999. *The Advanced Technology Program: Challenges and Opportunities*. Washington, DC: National Academy Press.
National Research Council. 1999. *Funding a Revolution: Government Support for Computing Research*. Washington, DC: National Academy Press.
National Research Council. 1999. *Industry-Laboratory Partnerships: A Review of the Sandia Science and Technology Park Initiative*. Washington, DC: National Academy Press.
National Research Council. 1999. *New Vistas in Transatlantic Science and Technology Cooperation*. Washington, DC: National Academy Press.
National Research Council. 1999. *The Small Business Innovation Research Program: Challenges and Opportunities*. Washington, DC: National Academy Press.
National Research Council. 2000. *The Small Business Innovation Research Program: An Assessment of the Department of Defense Fast Track Initiative*. Washington, DC: National Academy Press.
National Research Council. 2000. *U.S. Industry in 2000: Studies in Competitive Performance*. Washington, DC: National Academy Press.
National Research Council. 2001. *The Advanced Technology Program: Assessing Outcomes*. Washington, DC: National Academy Press.
National Research Council. 2001. *Attracting Science and Mathematics Ph.D.s to Secondary School Education*. Washington, DC: National Academy Press.
National Research Council. 2001. *Building a Workforce for the Information Economy*. Washington, DC: National Academy Press.
National Research Council. 2001. *Capitalizing on New Needs and New Opportunities: Government-Industry Partnerships in Biotechnology and Information Technologies*. Washington, DC: National Academy Press.
National Research Council. 2001. *A Review of the New Initiatives at the NASA Ames Research Center*. Washington, DC: National Academy Press.
National Research Council. 2001. *Trends in Federal Support of Research and Graduate Education*. Washington, DC: National Academy Press.
National Research Council. 2002. *Government-Industry Partnerships for the Development of New Technologies: Summary Report*. Washington, DC: The National Academies Press.
National Research Council. 2002. *Making the Nation Safer: The Role of Science and Technology in Countering Terrorism*. Washington, DC: The National Academies Press.
National Research Council. 2002. *Measuring and Sustaining the New Economy*. Washington, DC: National Academy Press.

National Research Council. 2002. *Partnerships for Solid-State Lighting.* Washington, DC: The National Academies Press.
National Research Council. 2004. *An Assessment of the Small Business Innovation Research Program: Project Methodology.* Washington, DC: The National Academies Press.
National Research Council. 2004. *Productivity and Cyclicality in Semiconductors: Trends, Implications, and Questions.* Washington, DC: The National Academies Press.
National Research Council. 2004. SBIR—Program Diversity and Assessment Challenges: Report of a Symposium. Washington, DC: The National Academies Press.
National Research Council. 2004. *The Small Business Innovation Research Program: Program Diversity and Assessment Challenges.* Washington, DC: The National Academies Press.
National Research Council. 2006. *Beyond Bias and Barriers: Fulfilling the Potential of Women in Academic Science and Engineering.* Washington, DC: The National Academies Press.
National Research Council. 2006. *Deconstructing the Computer.* Washington, DC: The National Academies Press.
National Research Council. 2006. *Software, Growth, and the Future of the U.S. Economy.* Washington, DC: The National Academies Press.
National Research Council. 2006. *The Telecommunications Challenge: Changing Technologies and Evolving Policies.* Washington, DC: The National Academies Press.
National Research Council. 2007. *Enhancing Productivity Growth in the Information Age: Measuring and Sustaining the New Economy.* Washington, DC: The National Academies Press.
National Research Council. 2007. *India's Changing Innovation System: Achievements, Challenges, and Opportunities for Cooperation.* Washington, DC: The National Academies Press.
National Research Council. 2007. *Innovation Policies for the 21st Century.* Washington, DC: The National Academies Press.
National Research Council. 2007. *SBIR and the Phase III Challenge of Commercialization.* Washington, DC: The National Academies Press.
National Research Council. 2008. *An Assessment of the SBIR Program at the Department of Defense.* Washington, DC: The National Academies Press.
National Research Council. 2008. *An Assessment of the SBIR Program at the Department of Energy.* Washington, DC: The National Academies Press.
National Research Council. 2008. *An Assessment of the SBIR Program at the National Science Foundation.* Washington, DC: The National Academies Press.
National Research Council. 2009. *An Assessment of the SBIR Program at the Department of Defense.* Washington, DC: The National Academies Press.

National Research Council. 2009. *An Assessment of the SBIR Program at the National Aeronautics and Space Administration.* Washington, DC: The National Academies Press.

National Research Council. 2009. *An Assessment of the SBIR Program at the National Institutes of Health.* Washington, DC: The National Academies Press.

National Research Council. 2009. *Revisiting the Department of Defense SBIR Fast Track Initiative*. Washington, DC: The National Academies Press.

National Research Council. 2009. *Venture Capital and the NIH SBIR Program*. Washington, DC: The National Academies Press.

National Research Council. 2010. *Managing University Intellectual Property in the Public Interest*. Washington, DC: The National Academies Press.

National Research Council. 2011. *Building the 21st Century: U.S.-China Cooperation on Science, Technology, and Innovation*. Washington, DC: The National Academies Press.

National Research Council. 2011. *Expanding Underrepresented Minority Participation: America's Science and Technology Talent at the Crossroads*. Washington, DC: The National Academies Press.

National Research Council. 2011. *Growing Innovation Clusters for American Prosperity*. Washington, DC: The National Academies Press.

National Research Council. 2011. *The Future of Photovoltaics Manufacturing in the United States.* Washington, DC: The National Academies Press.

National Research Council. 2012. *Building Hawaii's Innovation Economy*. Washington, DC: The National Academies Press.

National Research Council. 2012. *Building the Arkansas Innovation Economy*. Washington, DC: The National Academies Press.

National Research Council. 2012. *Building the U.S. Battery Industry for Electric-Drive Vehicles: Progress, Challenges, and Opportunities*. Washington, DC: The National Academies Press.

National Research Council. 2012. *Clustering for 21st Century Prosperity*. Washington, DC: The National Academies Press.

National Research Council. 2012. *Meeting Global Challenges: German-U.S. Innovation Policy*. Washington, DC: The National Academies Press.

National Research Council. 2012. *Rising to the Challenge: U.S. Innovation Policy for the Global Economy*. Washington, DC: The National Academies Press.

National Research Council. 2013. *Building the Illinois Innovation Economy: Summary of a Symposium*. Washington, DC: The National Academies Press.

National Research Council. 2013. *Building the Ohio Innovation Economy: Summary of a Symposium*. Washington, DC: The National Academies Press.

National Research Council. 2013. *Competing in the 21st Century: Best Practice in State and Regional Innovation Initiatives*. Washington, DC: The National Academies Press.

National Research Council. 2013. *Strengthening American Manufacturing: The Role of the Manufacturing Extension Partnership—Summary of a Symposium*. Washington, DC: The National Academies Press.

National Research Council. 2014. *SBIR at the Department of Defense.* Washington, DC: The National Academies Press.

National Research Council. 2014. *The Flexible Electronics Opportunity.* Washington, DC: The National Academies Press.

National Science Board. 2005. *Science and Engineering Indicators 2005.* Arlington, VA: National Science Foundation.

National Science Board. 2006. *Science and Engineering Indicators 2006.* Arlington, VA: National Science Foundation.

National Science Board. 2014. *Science and Engineering Indicators 2014.* Arlington, VA: National Science Foundation.

National Science Board. 2015. *Revisiting the STEM Workforce: A Companion to Science and Engineering Indicators 2014*. Arlington, VA: National Science Foundation.

National Science Foundation. 1999. *1999 SBIR/STTR Phase I Program Solicitation and Phase II Instruction Guide*. Arlington, VA: National Science Foundation.

National Science Foundation. 2004. *Federal R&D Funding by Budget Function: Fiscal Years 2003-2005 (historical tables).* NSF 05-303. Arlington, VA: National Science Foundation.

National Science Foundation. 2006. "SBIR/STTR Phase II Grantee Conference, Book of Abstracts." Office of Industrial Innovation. May 18-20, 2006. Louisville, Kentucky.

National Science Foundation. "Committee of Visitors Reports and Annual Updates." http://www.nsf.gov/eng/general/cov/.

National Science Foundation. "Emerging Technologies." http://www.nsf.gov/eng/sbir/eo.jsp.

National Science Foundation. "Guidance for Reviewers." http://www.eng.nsf.gov/sbir/peer_review.htm.

National Science Foundation. "National Science Foundation at a Glance." http://www.nsf.gov/about.

National Science Foundation. National Science Foundation Manual 14, *NSF Conflicts of Interest and Standards of Ethical Conduct.* http://www.eng.nsf.gov/sbir/COI_Form.doc.

National Science Foundation. 2006. "SBIR/STTR Phase II Grantee Conference, Book of Abstracts." Office of Industrial Innovation. May 18-20, 2006, Louisville, Kentucky.

National Science Foundation. 2006. "News Items from the Past Year." Press Release. April 10.

National Science Foundation, Office of Industrial Innovation. 2005. Draft Strategic Plan. June 2.

National Science Foundation, Office of Legislative and Public Affairs. 2003. SBIR Success Story from News Tip. Web's "Best Meta-Search Engine." March 20.

Nelson, R. R. 1982. *Government and Technological Progress.* New York: Pergamon.

Nelson, R. R. 1986. "Institutions Supporting Technical Advances in Industry." *American Economic Review, Papers and Proceedings* 76(2):188.

Nelson, R. R., ed. 1993. *National Innovation System: A Comparative Study.* New York: Oxford University Press.

O'Brien, W. 2013. "March-in Rights Under the Bayh-Dole Act: The NIH's Paper Tiger?" *Seton Hall Law Review* 30:1403.

Office of Management and Budget. 2004. "What Constitutes Strong Evidence of Program Effectiveness." http://www.whitehouse.gov/omb/part/2004_program_eval.pdf.

Office of the President. 1990. *U.S. Technology Policy.* Washington, DC: Executive Office of the President.

Organisation for Economic Co-operation and Development. 1982. *Innovation in Small and Medium Firms.* Paris: Organisation for Economic Co-operation and Development.

Organisation for Economic Co-operation and Development. 1995. *Venture Capital in OECD Countries.* Paris: Organisation for Economic Co-operation and Development.

Organisation for Economic Co-operation and Development. 1997. *Small Business Job Creation and Growth: Facts, Obstacles, and Best Practices.* Paris: Organisation for Economic Co-operation and Development.

Organisation for Economic Co-operation and Development. 1998. *Technology, Productivity and Job Creation: Toward Best Policy Practice.* Paris: Organisation for Economic Co-operation and Development.

Organisation for Economic Co-operation and Development. 2006. "Evaluation of SME Policies and Programs: Draft OECD Handbook." *OECD Handbook.* CFE/SME 17. Paris: Organisation for Economic Co-operation and Development.

Perko, J. S., and F. Narin. 1997. "The Transfer of Public Science to Patented Technology: A Case Study in Agricultural Science." *Journal of Technology Transfer* 22(3):65-72.

Perret, G. 1989. *A Country Made by War: From the Revolution to Vietnam—The Story of America's Rise to Power.* New York: Random House.

Poland, C. A., R. Duffin, I. Kinloch, A. Maynard, W. A. H. Wallace, A. Seaton, V. Stone, and S. Brown. 2008. "Carbon Nanotubes Introduced into the Abdominal Cavity of Mice Show Asbestos-Like Pathogenicity in a Pilot Study." *Nature Nanotechnology* 3(7):423.

Porter, Michael E. 1998. "Clusters and Competition: New Agendas for Government and Institutions." In Michael E. Porter, ed. *On Competition.* Boston, MA: Harvard Business School Press.

Powell, W. W., and P. Brantley. 1992. "Competitive Cooperation in Biotechnology: Learning Through Networks?" In *Networks and Organizations: Structure, Form and Action.* N. Nohria and R. G. Eccles, eds. Boston, MA: Harvard Business School Press. Pp. 366-394.

Price Waterhouse. 1985. *Survey of Small High-tech Businesses Shows Federal SBIR Awards Spurring Job Growth, Commercial Sales.* Washington, DC: Small Business High Technology Institute.

Reid, G. C., and J. A. Smith. 2007. *Risk Appraisal and Venture Capital in High Technology New Ventures.* New York: Routledge.

Roberts, E. B. 1968. "Entrepreneurship and Technology." *Research Management* (July):249-266.

Rogelberg, S., C. Spitzmüeller, I. Little, and S. Reeve. 2006. "Understanding Response Behavior to an Online Special Survey Topics Organizational Satisfaction Survey." *Personnel Psychology* 59:903-923.

Romer, P. 1990. "Endogenous Technological Change." *Journal of Political Economy* 98:71-102.

Rosa, P., and A. Dawson. 2006. "Gender and the Commercialization of University Science: Academic Founders of Spinout Companies." *Entrepreneurship & Regional Developmen*t 18(4):341-366. July.

Rosenberg, N. 1969. "The Direction of Technological Change: Inducement Mechanisms and Focusing Devices." *Economic Development and Cultural Change*, 18:1-24.

Rosenbloom, R., and W. Spencer. 1996. *Engines of Innovation: U.S. Industrial Research at the End of an Era.* Boston, MA: Harvard Business School Press.

Roy, A. S. 2012. "Stifling New Cures: The True Cost of Lengthy Clinical Drug Trials." FDP Project Report 5. Manhattan Institute. April.

Rubenstein, A. H. 1958. *Problems Financing New Research-Based Enterprises in New England.* Boston, MA: Federal Reserve Bank.

Ruegg, R., and P. Thomas. 2007. *Linkages from DoE's Vehicle Technologies R&D in Advanced Energy Storage to Hybrid Electric Vehicles, Plug-in Hybrid Electric Vehicles, and Electric Vehicles.* Washington, DC: U.S. Department of Energy/Office of Energy Efficiency and Renewable Energy.

Sahlman, W. A. 1990. "The Structure and Governance of Venture Capital Organizations." *Journal of Financial Economics* 27:473-521.

Saxenian, A. 1994. *Regional Advantage: Culture and Competition in Silicon Valley and Route 128.* Cambridge, MA: Harvard University Press.

Schacht, W.H. 2008. "The Small Business Innovation Research (SBIR) program: Reauthorization efforts," Congressional Research Service, Library of Congress.

Schell, J. K., and N. Berente. 2014. "Avoiding the Valley of Death: A Cross-Case Analysis of SBIR Innovation Processes." *Academy of Management Proceedings* 2014(1):16828.

Scherer, F. M. 1970. *Industrial Market Structure and Economic Performance.* New York: Rand McNally College Publishing.

Schumpeter, J. 1950. *Capitalism, Socialism, and Democracy.* New York: Harper and Row.

Scotchmer, S. 2004. *Innovation and Incentives.* Cambridge MA: MIT Press.

Scott, J. T. 1998. "Financing and Leveraging Public/Private Partnerships: The Hurdle-Lowering Auction." *STI Review* 23:67-84.

Scott, J. T. 2000. "An Assessment of the Small Business Innovation Research Program in New England: Fast Track Compared with Non-Fast Track." In National Research Council. *The Small Business Innovation Research Program: An Assessment of the Department of Defense Fast Track Initiative.* Washington, DC: National Academy Press.

Sheehan, K. 2001. "E-mail Survey Response Rates: A Review." *Journal of Computer Mediated Communication* 6(2).

Siegel, D., D. Waldman, and A. Link. 2004. "Toward a Model of the Effective Transfer of Scientific Knowledge from Academicians to Practitioners: Qualitative Evidence from the Commercialization of University Technologies." *Journal of Engineering and Technology Management* 21(1-2).

Silverman, I. M., J. M. Dawicki-McKenna, D. W. Frederick, C. Bialas, J. R. Remsberg, N. L. Yohn, N. Sekulic, A. B. Reitz, and D. M. Gross. 2015. "Evaluating the Success of the Small Business Innovation Research (SBIR) Program: Impact on Biotechnology Companies in Pennsylvania." *Technology* 2(1):5.

Silverstein, S. C., H. H. Garrison, and S. J. Heinig. 1995. "A Few Basic Economic Facts about Research in the Medical and Related Life Sciences." *FASEB* 9:833-840.

Sohl, J. 2014. "The Angel Investor Market in 2013: A Return to Seed Investing." *Center for Venture Research* April 30.

Sohl, J., J. Freear, and W. E. Wetzel Jr. 2002. "Angles on Angels: Financing Technology-Based Ventures—An Historical Perspective." *Venture Capital: An International Journal of Entrepreneurial Finance* 4(4).

Solow, R. S. 1957. "Technical Change and the Aggregate Production Function." *Review of Economics and Statistics* 39:312-320.

Specht, D. C. 2009." Recent SBIR Extension Debate Reveals Venture Capital Influence." *Procurement Law* 45(2009):1.

Stam, E., and K. Wennberg. 2009. "The Roles of R&D in New Firm Growth." *Small Business Economics* 33:77-89.

Stiglitz, J. E., and A. Weiss. 1981. "Credit Rationing in Markets with Incomplete Information." *American Economic Review* 71:393-409.

Stokes, D. E. 1997. *Pasteur's Quadrant: Basic Science and Technological Innovation.* Washington, DC: The Brookings Institution.

Stowsky, J. 1996. "Politics and Policy: The Technology Reinvestment Program and the Dilemmas of Dual Use." Mimeo. University of California.

Tassey, G. 1997. *The Economics of R&D Policy.* Westport, CT: Quorum Books.

Thurber, K. J. 2011. *Big Wave Surfing*. Edina, MN: Beaver Pond Press.

Tibbetts, R. 1997. "The Role of Small Firms in Developing and Commercializing New Scientific Instrumentation: Lessons from the U.S. Small Business Innovation Research Program." In *Equipping Science for the 21st Century.* J. Irvine, B. Martin, D. Griffiths, and R. Gathier, eds. Cheltenham UK: Edward Elgar Press.

Tirman, J. 1984. *The Militarization of High Technology.* Cambridge, MA: Ballinger.

Tyson, L., T. Petrin, and H. Rogers. 1994. "Promoting Entrepreneurship in Eastern Europe." *Small Business Economics* 6:165-184.

University of New Hampshire Center for Venture Research. 2007. *The Angel Market in 2006.*

U.S. Congress. House Committee on Science, Space, and Technology. 1992. *SBIR and Commercialization: Hearing Before the Subcommittee on Technology and Competitiveness of the House Committee on Science, Space, and Technology, on the Small Business Innovation Research [SBIR] Program.* Testimony of James A. Block, President of Creare, Inc. Pp. 356-361.

U.S. Congress. House Committee on Science, Space, and Technology. 1992. *The Small Business Research and Development Enhancement Act of 1992.* House Report (Rept. 102-554) Part I (Committee on Small Business).

U.S. Congress. House Committee on Science, Space, and Technology. 1998. *Unlocking Our Future: Toward a New National Science Policy: A Report to Congress by the House Committee on Science, Space, and Technology.* Washington, DC: Government Printing Office. http://www.access.gpo.gov/congress/house/science/cp105-b/science105b.pdf.

U.S. Congress. House Committee on Small Business. Subcommittee on Workforce, Empowerment, and Government Programs. 2005. *The Small Business Innovation Research Program: Opening Doors to New Technology.* Testimony by Joseph Hennessey. 109th Cong., 1st sess., November 8.

U.S. Congress. House Committee on Science, Space, and Technology. Subcommittee on Technology and Innovation. 2007. Hearing on "Small Business Innovation Research Authorization on the 25th Program Anniversary." Testimony by Robert Schmidt. April 26.

U.S. Congress. Senate Committee on Small Business. 1999. *Small Business Innovation Research (SBIR) Program.* Senate Report 106-330. August 4, 1999. Washington, DC: U.S. Government Printing Office.

U.S. Congress. Senate Committee on Small Business. 1981. Small Business Research Act of 1981. S.R. 194, 97th Congress.

U.S. Congress. Senate Committee on Small Business. 2006. *Strengthening the Participation of Small Businesses in Federal Contracting and Innovation Research Programs.* Testimony by Michael Squillante. 109th Cong., 2nd sess., July 12.

U.S. Congressional Budget Office. 1985. *Federal Financial Support for High-Technology Industries.* Washington, DC: U.S. Congressional Budget Office.

U.S. Department of Education. 2005. "Scientifically-Based Evaluation Methods: Notice of Final Priority." *Federal Register* 70(15):3586-3589.

U.S. Food and Drug Administration. 1981. Protecting Human Subjects: Untrue Statements in Application. 21 C.F.R. §314.12.

U.S .Food and Drug Administration. Critical Path Initiative." htttp://www-.fda.gov/oc/initiatives/criticalpath/.

U.S. General Accounting Office. 1987. *Federal Research: Small Business Innovation Research Participants Give Program High Marks.* Washington, DC: U.S. General Accounting Office.

U.S. General Accounting Office. 1989. *Federal Research: Assessment of Small Business Innovation Research Program.* Washington, DC: U.S. General Accounting Office.

U.S. General Accounting Office. 1992. *Federal Research: Small Business Innovation Research Program Shows Success But Can Be Strengthened.* RCED–92–32. Washington, DC: U.S. General Accounting Office.

U.S. General Accounting Office. 1997. *Federal Research: DoD's Small Business Innovation Research Program.* RCED–97–122, Washington, DC: U.S. General Accounting Office.

U.S. General Accounting Office. 1998. *Federal Research: Observations on the Small Business Innovation Research Program.* RCED–98–132. Washington, DC: U.S. General Accounting Office.

U.S. General Accounting Office. 1999. *Federal Research: Evaluations of Small Business Innovation Research Can Be Strengthened.* RCED–99–114, Washington, DC: U.S. General Accounting Office.

U.S. General Accounting Office. 1999. *Federal Research: Evaluations of Small Business Innovation Research Can Be Strengthened.* T-RCED–99–198, Washington, DC: U.S. General Accounting Office.

U.S. Government Accountability Office. 2006. *Small Business Innovation Research: Agencies Need to Strengthen Efforts to Improve the Completeness, Consistency, and Accuracy of Awards Data.* GAO-07-38. Washington, DC: U.S. Government Accountability Office.

U.S. Government Accountability Office. 2006. *Small Business Innovation Research: Information on Awards Made by NIH and DoD in Fiscal years 2001-2004.* GAO-06-565. Washington, DC: U.S. Government Accountability Office.

U.S. Public Law 106-554, Appendix I–H.R. 5667—Section 108.

U.S. Small Business Administration. 1992. *Results of Three-Year Commercialization Study of the SBIR Program.* Washington, DC: U.S. Government Printing Office.

U.S. Small Business Administration. 1994. *Small Business Innovation Development Act: Tenth-Year Results.* Washington, DC: U.S. Government Printing Office.

U.S. Small Business Administration. 1998. *An Analysis of the Distribution of SBIR Awards by States, 1983-1996*. Washington, DC: Small Business Administration.
U.S. Small Business Administration. 2003. "Small Business by the Numbers." SBA Office of Advocacy. May.
U.S. Small Business Administration. 2006. "Frequently Asked Questions." http://www.sba.gov/advo/stats/sbfaq.pdf.
U.S. Small Business Administration. 2006. "Small Business by the Numbers." SBA Office of Advocacy. May.
U.S. Small Business Administration. 2012. "SBIR Policy Directive." October 18.
Venture Economics. 1988. *Exiting Venture Capital Investments.* Wellesley, MA: Venture -Economics.
Venture Economics. 1996. "Special Report: Rose-Colored Asset Class." *Venture Capital Journal* 36 (July):32-34.
VentureOne. 1997. *National Venture Capital Association 1996 Annual Report*. San Francisco: VentureOne.
Wallsten, S. J. 1996. "The Small Business Innovation Research Program: Encouraging Technological Innovation and Commercialization in Small Firms." Unpublished working paper. Stanford University.
Wallsten, S. J. 1998. "Rethinking the Small Business Innovation Research Program." In *Investing In Innovation.* L. M. Branscomb and J. Keller, eds., Cambridge, MA: MIT Press.
Warner, E. E. 2014. "Factor Analysis of Disruptive Technology Approaches and Company Demographics in Defense SBIR Phase 1 Competition." Ph.D. Dissertation. Capella University.
Washington Technology. 2007. "Top 100 Federal Prime Contractors: 2004." May 14.
Weiss, S. 2006. "The Private Equity Continuum." Presentation at the Executive Seminar on Angel Funding, University of California at Riverside, December 8-9, Palm Springs, CA.
Whalen, P. S., S. S. Holloway, and I. D. Parkman. 2015. "Navigating the 'Valley of Death': an Investigation of Which Marketing Competencies Lead Toward Successful Technology Commercialization." In *Proceedings of the 2008 Academy of Marketing Science (AMS) Annual Conference*. Springer International Publishing. Page 184.
Yu, M-F., O. Lourie, M. J. Dyer, K. Moloni, T. F. Kelly, and R. S. Ruoff. 2000. "Strength and Breaking Mechanism of Multiwalled Carbon Nanotubes Under Tensile Load." *Science* 287(5453):637-640.